The Complete Book of Handcrafted Paper

Also by Marna Elyea Kern

An Introduction to Breadcraft

The Complete Book of Handcrafted Paper

MARNA ELYEA KERN

Coward, McCann & Geoghegan, Inc.
New York

Library of Congress Cataloging in Publication Data

Kern, Marna Elyea.
    The complete book of handcrafted paper.

    1. Paper, Handmade. I. Title.
ISBN 0-698-10989-9

Printed in the United States of America

To John & Bunny Kern, my second family,
and to Grandma Beth Elyea, with love.

Contents

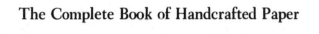

The Complete Book of Handcrafted Paper

Introduction

Why make paper by hand when you can go out to any one of a dozen different stores and buy some?

Because there is paper and there is paper.

The machine-made paper you can buy at a store is uniform in size, uniform in color and weight and perfect for a wide variety of uses—note taking, list making and so on.

Handmade paper is unique. No two sheets are exactly alike and no papermaker would want them to be. The beauty of handcrafted paper is in the fact that it does not look like paper made by a machine. And this is where you, the papermaker, come in.

You decide the color, weight, thickness, texture and size of the sheet of paper. You make all these decisions because you are in charge of the complete papermaking process from start to finish; this is the fun of making handcrafted paper. You get to create something from nothing and when you are done, you have a unique product—the result of your own creative skill.

Many different people are discovering handmade paper and for different reasons. Adults and children who are interested in crafts are drawn to papermaking because it is so versatile and so

11

inexpensive. Few tools are needed for this hobby and the raw material comes free in the mail. With a minimal investment of money, the craftsperson can produce distinctive stationery, attractive placemats and bookplates, plus other items to give as gifts, to sell or to keep. Directions for making these items, as well as others, are given in Chapter 9.

Hobbyist bookbinders often take up papermaking as a second craft to supply the beautiful endpapers for special bindings and the handmade paper for their books. French handmade endpapers are available at several dollars a sheet, but the colors and tones of these imported papers can be duplicated at home. The hobby papermaker can even create lovely marbled sheets of paper. (Chapter 7 shows how.)

Craftspeople who are looking for a part-time income will find many possibilities in papermaking. There are few hand papermakers in this country and the European mills often do not accept small lot orders. The result is that there are few suppliers of fine, handmade paper. Chapter 10 describes the markets for handcrafted paper and how the beginning papermaker can sell his or her products. The chapter also tells how to set up a home paper mill.

Artists are drawn to papermaking for many reasons. Printmakers especially enjoy working with handmade paper, but it is also appreciated by artists who work in watercolor, pastels and even oils.

Some artists like to control the entire creative process, including making the paper on which their work is done. Some artists feel that the paper itself is the medium. It can be shaped, embossed, dyed and molded; and many contemporary artists are experimenting in these areas. Other artists are unable to locate a paper with a special texture, color or shape, so they turn to making their own.

Artists also appreciate the permanence and durability of 100 percent rag handmade paper as well as another, less easily defined characteristic: personality. A sheet of handmade paper responds to watercolor, pastels or acrylics as no machine-made paper can.

Teachers, camp leaders and scouting leaders who are looking for a special project find papermaking an excellent choice for art class, history class or a science fair. In fact, Boy Scouts can even earn a merit badge in pulp and paper by learning about the modern paper industry and by making their own paper by hand.

Using simple methods, inexpensive equipment and recycled paper, even elementary school children can create beautiful sheets of handcrafted paper while they learn about the history of papermaking at the same time. Paper has been made successfully by second graders and the craft is versatile enough to appeal to students of all grade levels. The Embroidery Hoop Mold, described in Chapter 3, is so inexpensive that each student can work individually.

People interested in the value of recycling find papermaking an economical and also ecologically sound hobby. It not only reuses wastepaper and junk mail, but the craft also saves on the cost of buying stationery, notes, greeting cards and art paper.

Fund raisers, church bazaar sponsors and other groups find handmade paper a unique and very salable item at their money-raising events. Stationery made from handcrafted paper has a charm and appeal all its own. Chapter 9 tells how to make envelopes and how to package stationery for sale. The chapter also describes how to create note and greeting cards using simple printing techniques. Handmade paper products are particularly good money-makers because the raw material—junk mail—is free.

Making paper by hand is a "new" craft over 2,000 years old with many exciting possibilities. Discover some of these possibilities yourself in the next few chapters.

1

The History of Papermaking

History sections are often included in craft books to give the reader a feeling for the heritage of the craft. With papermaking, though, the history of the craft *is* the craft. Hand papermaking has remained virtually unchanged for some 2,000 years. The papermaker of today uses the same techniques as the Chinese in 105 A.D. or as the artisan during the Middle Ages or as the Colonial American paper-mill worker. When a modern papermaker dips that first sheet of paper, he or she becomes a member of a centuries-old group of craftspeople.

It all began when someone noticed that macerated fibers from bark, cloth or straw clung together when they were picked up out of water on a piece of woven cloth. It is believed that the earliest paper mold consisted of a square or rectangular box without a top or bottom. Into this box was fitted a screen, at first made of woven cloth, later of strips of bamboo. Many scholars feel that the earliest method for making paper was to pour the pulp and water into the boxlike mold, stir the solution briefly, then lift the screen out with a newly formed sheet of paper attached. The paper sheet was probably dried on the screen in the sun for 4 to 6 hours.

The actual credit for inventing paper is given to a court official of China named Ts'ai Lun who in 105 A.D. formed paper from the fibers of the mulberry tree by first shredding the bark, then mixing it with scraps of linen and hemp. Ts'ai Lun then beat this mixture into a pulp and poured it into a bamboo mold fitted with a cloth screen.

Of course, long before 105 A.D. civilizations had been recording information on paperlike substances. The Egyptians were making papyrus as early as 3100 B.C. However, papyrus is not considered true paper because it is not made from macerated fibers that form a matted sheet on a screen suspended in water. Rather, the reedlike papyrus plant is separated into thin strips and these are built up into layers in a laminating process. Chapter 12 describes the making of papyrus in detail with directions for making a modern version using readily obtainable plants.

Other paperlike materials were developed around the world. The *amatl* paper of the Aztecs was made from the bark of the fig tree. Tapa cloth, which was used for centuries by Pacific islanders, was made from the pounded bark of the breadfruit or mulberry trees. The North American Indians used birch bark as a writing material, and even today rice paper is still made in the Orient from the pith of a special tree. All these paper variations and others are described in Chapter 12.

When the Chinese first began making true paper, they boiled branches of the mulberry tree in a lye solution made from leached wood ashes. This solution removed the bark from the branches and separated the inner part of the bark into fibers. The Chinese then washed the fibers, strained them and worked them by hand into a pulp. Next they spread the pulp onto a table and beat it with a mallet until it was uniformly fine textured. They placed the prepared pulp into a tub with water and a sizing made from rice and special roots. Besides using fibers from the mulberry tree, the Chinese also used bamboo pith and, later, cotton imported from India.

One of the first early improvements in the method of making paper was the change from pouring the pulp into the mold to dipping the mold into a large vat of prepared pulp-and-water mixture. This new technique allowed the craftsman more control over the process, resulting in better sheets of paper.

Another important improvement in the papermaking process was a change in mold that allowed the paper sheet to be removed while it was still wet. In the Orient this mold consisted of a square or rectangular frame with supporting crossbars as well as a screen. The screen was made of thin bamboo strips set side by side and then stitched or laced together. These strips and stitches left impressions on the finished paper sheets known as "laid lines" (from the strips) and "chain lines" (from the stitches). These terms are still used today with reference to paper made on a laid mold.

After a sheet of paper was formed by dipping the mold and screen into a vat of pulp, the bamboo screen could easily be peeled off the paper. In this way, the same screen could be used many times without waiting for the paper sheets to dry, thus increasing the productivity of the papermaker. To keep the pulp from flowing over the sides of the mold, Chinese papermakers held loose sticks along the mold edges.

This was the state of papermaking when it spread west. In 751 A.D. the Arabs learned the secret of making paper from Chinese they captured at a paper mill in Samarkand. The Arabs made their paper from linen and it was several more centuries before papermakers in the West rediscovered the method of making paper from wood pulp.

The Arabs took their knowledge of papermaking to Europe where the first European mill was set up at Xativa, Spain, in 1151 A.D. Another early mill, which is still operating today, was established about 1276 at Fabriano, Italy. Afterward, papermaking spread to France around 1348 and later to Germany in 1390 and finally to England in 1494. By the end of the fifteenth century, paper had replaced parchment and vellum (both made from animal skins) in Europe and England, but it was not until printing with movable type was developed in the 1450s that there was any great demand for paper.

The method used by early paper mills required four main

17

Der Papyrer.

Ich brauch Hadern zu meiner Mül
Dran treibt mirs Rad deß waffers viel/
Daß mir die zschnitn Hadern nelt/
Das zeug wirt in waffer einquelt/
Drauß mach ich Pogn/auff dē filtz bring/
Durch preß das waffer darauß zwing.
Denn henck ichs auff/laß drucken wern/
Schneweiß vnd glatt / so hat mans gern.

F ij   Der

"The Papermaker," from a 1568 woodcut in *The Book of Trades* by Jost
Amman & Hans Sachs. The vatman forms a sheet of paper. Courtesy of
Dover Publications.

pieces of equipment: the vat, molds, felts and a press.

Old vats were about 5 feet wide and made of wood. Often they were old wine kegs cut in half. The bridge, a platform built across one edge of the vat, served as a shelf where the papermaker could rest molds with newly formed sheets so that the paper could drain.

Linen and cotton rags were used as the raw materials for papermaking until the 1800s. In the early days of papermaking there was no method of bleaching dyed rags (chlorine was not discovered until 1774) so even the highest-quality paper was cream colored or even grayish. This paper was made by laboriously hand-picking the best rags—those of white linen. Other grades of paper were often brownish in color or even speckled.

Rags to be used for paper were first allowed to ferment for 6 to 7 weeks to loosen the fibers. Next the rags were cut up and washed in an attempt to remove the yellow color that resulted from the fermentation process. This was not usually completely successful.

The next step in processing the rags was beating. The earliest method was to beat the rags between two stones, next a mortar and pestle were used and later a stamping machine. In the stamping machine a row of water-powered hammers beat the rags in lead- or iron-lined troughs. Often some of the hammers were equipped with spikes or iron teeth to fray the rags. Water was also used as part of the process. The running water served to wash the rags, replacing the earlier method of hand washing.

In the mid-1700s the Hollander beater was developed. This machine consisted of an oblong tank with a heavy wooden roll that was fitted with iron knives. The linen and cotton rags were added to the Hollander with water and they circulated around the tub. As the roll revolved against a metal or stone bedplate in the bottom of the tank, the rags were caught in between and frayed and split into fibers.

When the rags had been processed into pulp by a Hollander beater or a stamping machine, they were then added to the vat. To keep the pulp from settling to the bottom of the vat, the papermaker used a long pole to stir the mixture occasionally.

The mold used by European papermakers in the fifteenth century had already undergone some change from the Oriental version. Wire screens made of fine woven brass were permanently attached to the mold frame. Many different types of wood were used to make early molds, including bamboo, oak, pine, mahogany and fig. The size of the mold was limited to what a man could lift from the vat.

A boxlike structure, called the deckle, was constructed to fit snugly on top of the mold. This deckle kept the pulp from flowing over the edges of the mold during paper formation and it also created a relatively smooth edge to the paper sheet.

The actual method of making paper changed little in Europe from the fifteenth to the nineteenth centuries. In 1690 the first American paper mill was set up in Germantown, Pennsylvania, near Philadelphia, by William Rittenhouse. The mill produced 100 pounds of paper a day. American papermakers used the same general methods as outlined here:

After the vat had been filled with beaten pulp and water, the vatman took a mold and deckle and, holding them firmly together, dipped the unit into the pulp at an almost vertical angle until the mold was completely immersed in the pulp. Then the vatman turned the mold unit to the horizontal and brought it to the surface of the vat. Fibers clung to the screen of the mold, forming a sheet of paper.

Next the vatman shook the mold to remove excess water and also to strengthen the fiber bonds of the pulp. The shake was unique to each papermaker and was developed after much practice. After this step, the deckle was removed from the mold, leaving a clean-edged sheet of wet paper on the screen. A good vatman could make about 750 sheets of paper a day.

The mold with its new sheet of paper was then passed on to a second craftsman, the coucher, whose job it was to remove or couch (pronounced *coosh*) the moist sheet. First the coucher set the mold at an angle, allowing even more water to drain back into the vat. When the sheet of new paper had dried to just the right degree (learned only after much experience), the coucher then turned the mold over, dropping the unwrinkled sheet of wet paper onto a piece of felt slightly larger than the paper itself. The wet sheet of paper was then covered with another felt, ready for the next sheet to be couched.

When a pile of 144 sheets of paper had been stacked with felts, the whole stack was set in a press and the screw of the press turned down. This step removed excess water and matted the fibers of the new paper so that it could easily be removed from the felts.

The next step in the papermaking process involved yet another craftsman, the layman. He removed each sheet of paper from the felt and stacked it in a pile with other moist sheets. This new pile of sheets was then pressed, taken apart, rearranged and pressed again. The process was known as exchanging and it was repeated until the paper had developed the desired smooth finish and texture. Paper made before the sixteenth century is often rough and some scholars believe that exchanging was not a common practice before that time.

After the exchanging process was completed, the still-damp paper sheets were taken in spurs of four or five sheets and dried in the loft of the mill. The paper was kept in spurs of several sheets because a single sheet would wrinkle and curl if dried alone, but the weight of several sheets produced even drying in all. Paper spurs were hung either on wooden poles or pegs or placed over horsehair ropes that had been coated with beeswax. Some old paper shows the stain marks from this drying process.

After the paper was dry, it had to be "sized" if it was to be used for writing. Paper destined for the printer could be packed up and sent on its way without any additional sizing.

Sizing was made from scraps of animal hide, boiled until it became a gelatinous liquid. The dried sheets of paper were dipped into the warm sizing, then pressed to remove excess liquid and dried once again. So much damage was done to the paper during this step, resulting in torn and ripped sheets, that the sizing room came to be known as the "slaughterhouse."

Finishing was the final step in the papermaking process. The early method of finishing paper to produce a smooth surface

21

was to rub each sheet by hand with an agate, soapstone or flint. Hand polishing always gave the paper a streaked appearance. Later a mechanical glazing hammer replaced hand rubbing and produced a uniform, smooth surface on the paper sheet. After this last step, the paper was ready to use.

The ever-increasing demand for paper created by the invention of movable type and the birth of the printing industry resulted in rag shortages in both Europe and America. Shortages became so severe at times that paper mills even advertised for rags, and laws were passed to limit the size of newspapers, in an effort to conserve paper. During the American Revolution, paper was so scarce that soldiers were forced to tear up old books to make wadding for their guns, and official messages were sent to General Washington written on scraps and bits of paper.

Because of the worsening rag shortages, experimenters began looking for alternate sources of pulp. Cotton stalks, thistle stalks, cornhusks, burdock stalks, potatoes, aloe, willow, rush, and wasps' nests were among the materials used at one time or another. Early newsprint was made of straw, but it was of poor quality. Later, after 1774 when chlorine bleaching was introduced, hemp rope and burlap bags were used as pulp sources. The first book printed in Europe on paper made from materials other than rags was issued in 1784 in France. The paper was made from grass, lime tree bark and other plant fibers.

In the mid-1800s a German handweaver, Friedrich Gottlob Keller, discovered that pulverized wood could be used to make paper. He developed a machine for grinding the wood into fibers. However, paper made from wood pulp had many defects. Because the wood fibers were shorter than rag fibers, the paper produced from them was not as strong. Resins and other impurities also resulted in weakening and discoloration. Soon, though, English papermakers developed a method to remove these resins by boiling the wood pulp with soda ash.

In 1867 an American, B. C. Tilghman, developed the sulfite process to break down wood. In this process, the bark is first removed from logs, then the wood is chipped into small pieces that are boiled under pressure with sulfurous acid.

These new methods of obtaining and processing pulp for papermaking combined with an earlier invention—the papermaking machine—to revolutionize the paper industry. In 1799 Nicolas-Louis Robert, a clerk at a French paper mill, developed a moving-screen belt device that produced an unbroken sheet of wet paper without hand dipping. A London stationer, Henry Fourdrinier, brought the idea to England and the Fourdrinier paper machine was developed. In 1809, the cylinder paper machine was invented. Both developments brought mechanization to papermaking, but production was still limited by the shortage of rags. With Keller's discovery of wood pulp as a paper source a few decades later, the modern paper industry was born.

MODERN PAPERMAKING

Today the soda ash process is used to prepare groundwood pulp for newsprint while the sulfite process is used to make fine papers. The kraft or sulfate process is used to make strong paper for grocery bags and wrapping material.

What is paper and how does it actually form? Chemists have discovered that the bonding that develops as a sheet of paper dries is due to the interatomic forces of attraction between the cellulose fibers. These forces are known as van der Waals forces. The forces are neutralized in water which is why a wet sheet of untreated paper has little strength.

Fiber Sources

The main source of fiber for modern papermaking is pulped forest tree trunks. However, only a little more than half of the wood pulp used to make paper comes from trees cut specifically for papermaking. Some 23 percent of the wood pulp used in paper mills comes from waste materials such as sawdust and wood chips from sawmills and lumbering operations. Another

22 percent of the pulp used in modern papermaking comes from recycled wastepaper, while 2 percent is from nonwood sources like cotton and linen rags, which are obtained from garment-mill cuttings.

Wood Pulp. Spruce has a high percentage of cellulose and is a good pulp source while hemlock makes a coarse grade of paper. Fibers from softwood trees (most are evergreens) are longer and add strength to the paper while the fibers from hardwoods (deciduous trees) are shorter. These serve to fill in the sheet of paper and give it opacity and a smooth finish.

Trees destined for wood pulp are cut to prescribed lengths in the forest, then transported to the paper company where they are fed into a barker, which strips away the bark and cleans the wood. Next the wood goes to a chipper, where it is ground into small pieces.

There are two methods of preparing pulp. Groundwood pulp is produced by mechanically grinding the wood chips into fibers. It is not as strong as chemical pulp but it is useful in high-speed printing. Chemical wood pulp is made by cooking the wood chips with chemical solutions under pressure and at a high temperature. This dissolves the lignin (a cellulose-like substance which binds fibers and strengthens cell walls) and frees the cellulose fibers in the pulp. The pulp is then cleaned, screened and bleached, after which it is ready to be used for papermaking.

Rags. Rags are received in large bales at the paper mill where they must be sorted to remove debris and any synthetic materials that may be difficult to process. Today, as earlier, the best rags are of fine linen. After sorting, the rags are cut into small pieces and cooked in chemical solutions of lime and soda ash or caustic soda with wetting agents or detergents. This removes any oil, grease or fillers in the material. Next the rags are washed and then mechanically beaten to break down the fibers.

Very little paper is made from just one type of pulp, except for the finest-quality rag paper. Cotton and linen fibers are used to make rag paper, which can vary in rag content from 25 percent to 100 percent. Since rag fibers are finer and longer than wood pulp fibers, the resulting paper is stronger and more permanent. Rag paper is used for important documents where long life is a determining factor, such as stock certificates and bank notes, and it is also used for high-quality stationery.

Wastepaper. The paper industry recycles large amounts of wastepaper but there are limits to this recycling. Wastepaper cannot be used to make a higher-quality paper than the original paper product. In addition, wastepaper can only be recycled a certain number of times before the fibers break down into lengths that are too short for use.

While only a small percentage of the reclaimed wastepaper can be used to make printing-grade paper, the rest can be used to make coarser papers such as boxboard.

Wastepaper is received at the paper mill in bales and consists of out-of-date books and unsold magazines, newspapers, telephone directories, catalogs, legal papers and wastepapers from schools or offices.

As in the case of rags, the first step in preparing wastepaper for pulp is sorting. Next the wastepaper intended to become printing-grade paper must be de-inked, which is accomplished in a pulper, a tank with agitator blades that stir the recycled paper stock as it is washed with chemicals such as caustic soda together with hot water. The resulting pulp is then washed again and, depending on its intended use, sometimes bleached with calcium hypochlorite.

Beating
Whether the fiber comes from rags, wood pulp or recycled wastepaper, the next step in the preparation of pulp is beating. The Hollander beater or a variation on this design is still used in smaller mills. Large mills use a type of continuous refiner for altering the fiber in the pulp. These refiners consist of an open container with one or more rotating blades which split and fray the fibers during beating. Dyes, fillers and sizings can all be added to the pulp during the beating stage.

The Cylinder Paper Machine
Paper made on this machine is called mold made and

resembles handmade paper as closely as is possible for machine-made sheets. The cylinder machine consists of one or more cylinders covered with a woven-metal wire screen. The cylinder is partially immersed in a tub of pulp and water and revolves slowly in this tub. The pulp remains on the outside of the screen, thereby forming a wet sheet of paper. This paper sheet can be peeled off the mold by hand, in the case of heavy paper, or it can be passed on to presses and dryers as described below.

The cylinder machine can be used to make laminated sheets of paper consisting of several layers. Several wet sheets of paper are pressed together to form a laminate.

The first American cylinder machine was used in Wilmington, Delaware, in 1817.

The Fourdrinier Paper Machine

After the pulp has been prepared, it is fed into the Fourdrinier paper machine through the headbox. Here pumps spray a thin film of pulp onto a continuous fine-screen belt, called a wire. The pulp is matted together and water drains from the paper as it travels along the wire in a continuous sheet. The wire of the Fourdrinier is woven of synthetic-fiber cloth or specially annealed bronze or brass with a mesh of 55–85 strands per inch. Fine wire screens are used to manufacture cigarette paper and coarse screens are used to make paperboard and other heavy-duty paper products. The wire screen is removable and must be replaced periodically due to wear. The Fourdrinier wire can move at a rate of 200 feet per minute to 4,000 feet per minute and the width of the wire can range from 4 to 30 feet.

After the pulp has been sprayed onto the wire, the still-damp sheet of new paper travels through many different rollers, one of which is called the dandy roll. This roller is covered with wire cloth and serves to flatten the top surface of the paper sheet. Various patterns can also be given to the paper with the dandy roll: for woven paper, the roller makes a light mesh pattern on the sheet; for laid paper the dandy roll embosses the sheet with parallel, translucent lines, reminiscent of the laid

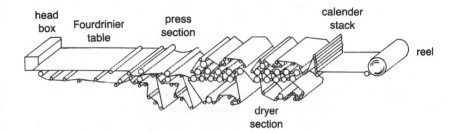

head box Fourdrinier table press section calender stack reel dryer section

The Fourdrinier paper machine.

lines from the bamboo strips of early hand molds. Watermarks can also be added to the paper by the dandy roll.

The last roller on the paper sheet's journey through this section of the Fourdrinier machine is the couch roll. Here suction removes water from the wet paper.

The next component of the machine to receive the paper sheet is the press section. More water is removed here and the paper fibers are compacted. Belts made of cotton, asbestos or plastic felt are used in the press section to absorb water from the new paper without crushing the fragile sheet. Like the wire, felts must be replaced periodically due to wear.

After the press section, the paper travels to the dryer section where forty to seventy steam-heated drying cylinders work to complete the drying of the paper sheet. Large amounts of water vapor are released into the atmosphere during this process: for every ton of paper dried on the machine, about two tons of water are evaporated.

Next the paper sheet may go through the calenders, heavy polished rollers that compact the paper sheet and give it a smooth finish.

Paper made on the Fourdrinier is produced in one long continuous sheet. At the end of the machine this sheet is wound onto a roller. Later the paper is rewound and cut to desired size.

Often, though, the paper is still not ready for use and it will be processed further to obtain special surfaces for different uses. Coatings are often applied to the paper in roll form to prepare the surface for special types of printing; for example, halftone printing. Clay, titanium dioxide and calcium carbonate are pigments that are often used to coat paper. To fix the coating to the paper sheet, adhesive binders such as starch, casein, latex and resins are used. Depending on the intended use of the paper, waterproofing materials—wax and polyethylene—can also be applied as coatings. Paper can be coated

The continuous sheet of paper is rolled onto a reel at the No. 7 paper machine, Union Camp, Savannah, Georgia. Photo by Susan P. Driscoll.

either on the Fourdrinier machine before being dried or later by another machine in a separate step.

Types of Paper Produced by Machine

Fine or Bond Paper. This group includes some of the highest-grade papers produced, with rag content varying from 25 percent to 100 percent. Fine stationery as well as legal documents, checks and paper money are all made from high-quality bond paper.

Paper to be used for writing or printing often contains white pigments or fillers to improve the brightness, opacity and smoothness of the sheet. Some of the pigments frequently used in commercial mills are clay (aluminum silicate), titanium dioxide, calcium carbonate (used mainly for magazine and printing papers), zinc oxide, zinc sulfide, talc and calcium sulfate.

Printing Paper. Book paper comes under this heading and it is usually made from wood pulp, though some lower-priced grades are made from recycled wastepaper. Uncoated book paper is used for catalogs and circulars while coated book paper, treated with sizings, fillers and dyes to obtain a uniformly smooth surface, is used for fine screen halftone printing.

Printing paper must be opaque so that there is little "show through" from one side of the sheet to the other. To accomplish this, white mineral pigments are often applied as a coating or are added to the pulp during beating.

Newsprint. Newsprint is made mainly from groundwood pulp. This paper receives printing well, but because of a high content of lignin, newsprint yellows quickly and becomes brittle with age.

Coarse Paper. These papers are used to wrap and package products. Kraft paper is a type of wrapping paper that is made of unbleached wood pulp. Wet-strength kraft paper is treated with special resins to retard wetting. Grocery bags, sacks for shipping sugar, flour, cement and sand are all made of coarse paper as are heavy envelopes, asphalting paper and gummed paper.

Sanitary Paper. In this category are toilet tissue, paper towels, napkins and facial tissue. Some are wet strength, some are not. Sanitary paper such as paper towels and napkins are often embossed with patterns while the paper is still slightly moist.

Paperboard. Usually made from wood pulp, straw, wastepaper or some combination, paperboard has many uses. Paper boxes, corrugated shipping cartons and cardboard are just a few.

Industrial Paper. Paper in this category is produced for special industrial uses. Some examples are electrical insulation paper, filter papers and abrasive papers.

Wet-Strength Paper. To counteract the normal tendency for the fiber bonds of paper to dissolve in water, organic resins are added to the pulp and these are then absorbed by the fibers. After the sheet of paper dries, the resins become insoluble, producing water-resistant bonds between the fibers. Some examples of wet-strength papers are: paper towels, napkins, some wrapping papers, and paper diapers.

Bible Paper. This category of paper is thin and lightweight yet very strong and opaque. It is used to make bibles, dictionaries and encyclopedias. Pigments like titanium dioxide and barium sulfate are often added.

Bristol. The term Bristol applies to stiff, heavy paper that is thicker than 0.006 inch. Combinations of wood pulp are used in the manufacture of Bristol and sizing is usually added to give some wet strength to the paper.

Art Papers. Fine-quality watercolor paper is made by hand and has a 100 percent rag content. However, watercolor paper can also be made on a cylinder paper machine, in which case it is called mold made. Most of the watercolor paper sold today is machine made and can vary in fiber content from all wood pulp to 100 percent rag. The weight of watercolor paper also varies, with some common weights being basis 90 lb., basis 140 lb., basis 200 lb. and basis 300 lb. These basis weights refer to the weight of one ream (500 sheets) of paper in a 22" × 30" sheet size.

Charcoal and pastel papers are mainly machine made and can also vary in fiber content from all wood pulp to 100 percent rag.

Bristol board is made by laminating several sheets of drawing paper together to increase the rigidity of the paper. Bristol board is available in two, three, four or five plies and a range of fiber contents too.

2

Setting Up

There are very few materials needed for making paper. In fact, you probably have most of the equipment on hand:

>   measuring cups and spoons
>   a sink, utility tub, bathtub or large washbasin
>   oven
>   iron
>   blender
>   plastic or enamel colander
>   large stainless steel or enamelware kettle (a canning
>   kettle is ideal)
>   a metal baking sheet

The only special piece of equipment used in this craft is the mold and Chapter 3 gives directions for making five different types.

Other helpful tools include:

>   a blunt knife
>   long-handled stainless steel or wooden spoons (for
>   stirring the pulp during boiling)

small stainless steel pan (for melting sizing)
wooden cutting board (for iron drying)

In addition to these tools, you will need the following supplies for your papermaking:

Couching Cloths. To remove newly formed sheets of paper from the mold, you will use sheets of fabric—cotton or muslin—cut into squares or rectangles 1" larger on all sides than your mold. To avoid the problem of unraveling, cut the couching cloths with pinking shears.

You can use fabric remnants you have on hand to make the couching clothes or cut up an old sheet or buy a yard or two of inexpensive material from a fabric store. If you are purchasing material, choose natural muslin or a white- or cream-colored cotton. Avoid synthetic and cotton blends and dyed materials.

If you have some patterned or colored fabric remnants around the house that you would like to use for couching cloths, wash them first in hot water to be sure that the dye is colorfast and will not bleed onto your new sheets of paper.

Wash couching cloths after each use. Since small pieces of pulp adhere to the surface of the material, if the same cloth is used without washing to couch another sheet of paper, these bits of pulp can cause problems, making holes, lumps or tears in the new paper. After washing, hang cloths on a line to dry or use the cool setting of your clothes dryer. Stack the cloths as soon as they are dry to avoid wrinkling. Couching cloths do not have to be perfectly smooth, but large creases and wrinkles could be troublesome. If the cloths come out of the dryer hopelessly wrinkled and if stacking them does not smooth out the material, then use an iron to press the cloths. It will save trouble later when you are couching new sheets of paper.

Felts. Just as in the earlier days of hand papermaking, sheets of felt are used to absorb excess moisture from the newly formed paper when it is stacked. Buy felt by the yard at fabric stores; it is much cheaper than precut squares of felt sold at hobby and crafts stores. The size of the felt should also be about 1" larger on all sides than your mold. Buy white- or cream-colored felt since the dyes used on felt are often not colorfast and can streak a new sheet of damp paper.

Some felt shrinks. Cut out two test pieces the same size and wash one, then dry it in the dryer. Compare sizes. If shrinkage has occurred, be sure to use cold water to wash the felt in the future or hand wash the squares, then line dry them.

Felt seldom unravels along the cut edges so regular scissors can be used for cutting the felt into squares or rectangles to fit your mold.

Towels, extra felt or sponges. To couch a new sheet of paper from the mold with a minimum of problems, you will be using an absorbent material to blot up excess water through the couching cloth. Old towels work fine. You will need two to three hand-size towels for a twenty-sheet batch of paper. If you do not have old towels around, buy inexpensive kitchen or bath towels at a discount store. Do not use good towels for this step because the dyes used with paper can bleed through and discolor the towel.

Instead of towels, you can also use additional felt squares. Double the felt for extra absorbency. Sponges can also be used, but are not as effective.

Other supplies for your papermaking:

household bleach
talcum powder
cornstarch
unflavored gelatin
spray starch

This is not a complete list. There are many other materials you can use to create different special effects in your paper that will be described in later chapters.

SOURCES OF PULP

You do not need to look far for the raw material to use for hand papermaking. Every day your mailman delivers something that can be turned into new sheets of paper. In a week, maybe two at the most, you will have accumulated enough junk mail to make hundreds of sheets of new paper.

When you recycle junk mail into new paper, you are using virtually the same methods that large commercial paper mills use, but on a smaller scale. Paper mills have been recycling wastepaper for many years now with the wastepaper coming from many different sources: offices, schools, magazine stands, newspaper offices and so on.

The first step in recycling paper for you or for a paper mill is sorting. Not everything you find in your mailbox or collect from other sources can be turned into handmade paper; so here are some tips on what to look for and what to avoid:

High-Quality Paper

The better the paper quality to start with, the better the new sheet of paper will be. When recycled, rag paper with a 25 percent or higher content of cotton fiber will create strong new sheets. Some artists recycle their 100 percent rag watercolor paper (see Chapter 9). Stationery-grade paper usually recycles into stronger sheets than pamphlets or brochures.

Save high-quality, white sheets of paper with cotton-fiber content and process these separately for your own top-grade paper.

Unprinted Paper

The less printing, typing, writing or other ink, the better. While most of the ink from printed paper will bleach out during the boiling stage of pulp preparation, "clean" white paper is still preferred for recycling. Therefore, envelopes are a prime candidate since there is so little printing in proportion to the amount of paper. Use envelopes addressed to you and also the envelopes included in many mail offers. The glue on the flaps does not matter—it will wash away in the boiling water. But do be sure to remove any cellophane window panels since these will not break down into pulp. Postage stamps do not break down well either.

Colored Paper

Colored paper is just as good as white paper for recycling. In fact, you can still make white paper using a small percentage of colored pulp added to a predominantly white batch. The resulting white will not be as bright, however, unless you add whiteners (Chapter 6 discusses this). A rainbow mixture of junk mail boils down to an attractive ivory color if you add some bleach.

If you want to make colored sheets of paper, you can start with colored junk mail and leave out the bleach during the boiling step. Look for brightly colored sheets of junk mail and sort these by color for separate processing. The pulp can be used to make pastel-colored sheets of new paper. (Or small bits of colored pulp can be added to white sheets for an interesting marbled effect, see Chapter 7.) If you are planning to dye the pulp, you can use any color paper at all.

Paper with Special Surfaces

Paper with a slick surface, like that used in catalogs, some advertising brochures and greeting cards, usually breaks down into a gray-colored pulp. Sometimes this is desirable. Do avoid papers that are wet strength, though. As mentioned in Chapter 1, these papers have been specially treated with resins to make them water resistant and they do not break down well at all during boiling.

Newsprint

Even though the large paper mills recycle newsprint, it is not a recommended source of pulp for the home papermaker. Newsprint makes a weak pulp that forms gray sheets of paper. It is not really worth the bother when there are so many other, better pulp sources readily available.

Unbleached Kraft Paper

Kraft paper, the kind used in grocery bags and wrapping paper, is tough and can add strength to your pulp. However, it also reduces the brightness of the finished sheet. Sometimes you may want both these effects; if not, add bleach during the boiling process.

Use kraft paper in moderation, however, since it takes a long time to break down during boiling and even a small quantity mixed with other pulp will add strength to new paper. Some papermakers like to add a bit of unbleached kraft pulp to other pulp because it gives an antique look to new sheets.

White or colored bags used by some stores are not made from kraft paper. They can be recycled and used like the unprinted white or colored paper described above.

Facial Tissue and Toilet Tissue

While you can add these papers to your pulp, they are much more expensive to use than the recycled wastepaper and junk mail. However, facial and toilet tissues that are not "wet-strength" can be used without boiling so they provide a convenient pulp for demonstrations or for use with school groups. Chapter 4 gives specific directions for using this type of pulp.

Gather paper for recycling as it comes in the mail and store in a paper bag, box or closet until you have a good quantity. You will need about thirty to forty sheets of paper or about forty envelopes to make a good-sized batch of pulp. If you are collecting certain colors of paper, it may take you longer than a few weeks to accumulate this amount. Ask friends to help.

Do not overlook other sources of pulp around your house. Stationery and envelopes with out-of-date addresses can always be recycled into new sheets of good-quality paper. Old checks can also be used but the paper may take longer to break down during boiling.

Nonpaper Pulp Sources

If you owned a beater, you could make 100 percent rag paper by recycling old clothes. Unfortunately, though, beaters are expensive. Chapter 10 discusses some alternatives to traditional beaters.

Even without a beater, though, you can add small amounts of fabric to your paper for special colors or textures and to add variety. Choose from the following sources:

Natural Fabrics. Fabrics such as wool, cotton, linen or muslin can all be used in small amounts with paper pulp. Chapter 4 tells how to process these materials in a home blender.

Synthetic Fabrics. Other good rag pulp sources are acrylic, rayon and Orlon remnants. Use these in combination with recycled paper pulp since synthetics do not make strong sheets of paper when used alone.

Yarn, String, Thread. Synthetic, wool or mohair yarn, cotton or jute string, and embroidery thread can all be cut up into small pieces and blended with other pulp. Or the fibers can be added to paper sheets in larger pieces.

Lint. The lowly stuff that accumulates in the clothes dryer's lint trap consists mostly of fibers. If you remove any hair, paper clips or other foreign materials, you can add this lint to a blenderful of recycled paper pulp. Look for colorful lint—the kind that collects after you dry a bright blanket or sweater.

Nonwoody Plants. Since all plants contain cellulose and since cellulose is the ingredient used to make paper, it would seem that paper of some sort could be produced from any plant. And this is true. Chapter 11 gives directions for making paper with marigolds, grass, shrubs and roses, among other plants.

3

The Mold

Early papermakers used molds made of oak, mahogany, bamboo or teak. Modern papermakers can use anything from a picture frame to an embroidery hoop to a custom-built wooden mold with a deckle. Your choice of mold will be determined by how much time you plan to spend making paper and which method—vat or pouring—you want to use. (Chapter 4 provides instruction in both methods.)

Mold size need not be limited to the size of your kitchen sink, either. Consider making paper in a large washbasin, utility tub or even the bathtub.

EMBROIDERY HOOP MOLD

Anyone can put this mold together and it has the advantage of creating unique oval or round sheets of paper. This design is a good choice when many simple, inexpensive molds are needed, as for large groups of people or for schools where each person wants to participate in papermaking. The mold is small enough so that it can be used with a dishpan serving as the vat.

41

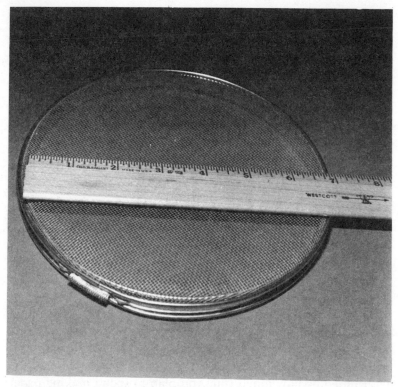

The completed Embroidery Hoop Mold, ready to make paper. Photo by Chris Miller.

Materials:

> a two-part wooden or metal embroidery hoop, oval or
>   round, any size
> fiberglass-mesh window screening
> varnish or polymer coating

If you are using a wooden embroidery hoop, waterproof it first by varnishing or by applying a coat of polymer. Let dry.

Set the embroidery hoop on top of the screening. Cut out the screen, leaving about 1″ to 2″ extra all around.

Open up the embroidery hoop. Place the cut section of screening on top of the inside hoop, then set the outside hoop on top and press into place. The screen should be pulled taut in the hoop. If it is not, repeat the process until the screen is smooth and taut. Trim off any excess screening.

The mold is now ready for use.

After repeated use, a metal embroidery hoop may rust slightly on the inside. To prevent this, always take the hoop apart after each papermaking session and wipe off the rims of the hoop, especially the insides. Air dry before storing.

The Picture Frame Mold consists of fiberglass-mesh screening stapled to the back of a picture frame. Photo by Chris Miller.

PICTURE FRAME MOLD

This mold is easy to make and some of the materials you need are probably already available around your house. Materials:

>wooden picture frame
>fiberglass-mesh screening
>staple gun, staples
>chalk
>varnish or polymer coating

Use a wooden picture frame of any convenient size, keeping in mind the size of the sink or tub you will use later. Apply varnish or a polymer coating to waterproof the frame and let dry.

Set the frame on top of the screening and mark off the frame size with chalk. Cut out the screen, leaving about 1" extra on all sides.

Turn the picture frame upside down and staple the screen along one edge. Place the staples close together so that no gaps are created between the frame and the screen when you pull the screen tight and staple it to the opposite side of the frame. Repeat for the two remaining sides. Trim off any excess screening.

The mold can now be used.

CANVAS-STRETCHER MOLD

This is an easy-to-make wooden mold with no nailing involved. Buy canvas-stretcher strips. These are available at most art supply stores in a wide range of lengths, so it is easy to create a custom-sized mold in either a rectangular or square shape. Materials:

>four canvas-stretcher strips (for a square mold, all four
>   strips the same size; for a rectangular mold, two strips
>   each of two different sizes)
>waterproof wood glue
>fiberglass-mesh screening

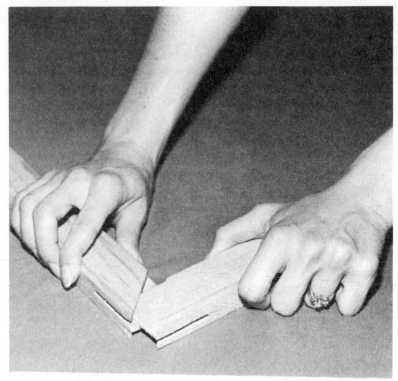

Canvas-stretcher Mold. Set the canvas-stretcher strips together to form the frame for this mold. Photo by Chris Miller.

staple gun, staples
varnish or polymer coating

Coat the ends of two stretcher strips with glue and set the strips together. (See illustration.) Repeat with the other two strips. Add glue to the remaining ends of these two L-shaped sections and fit them together to form a frame. Let dry.

Waterproof the frame with a coating of varnish or polymer sealer. Let dry.

Set the frame on top of a piece of fiberglass-mesh screening and cut the screen to fit. Staple the screen to one edge of the frame, placing the staples close together. Stretch the screen to the opposite side of the frame and staple here too. Repeat for the remaining sides.

45

EASILY CONSTRUCTED MOLD

Making this mold does not require a high degree of woodworking skill and it is also inexpensive to make. Materials:

1″ × 2″ trim
hammer, nails
sandpaper
fiberglass-mesh screening
staple gun, staples
chalk
varnish or polymer coating

Buy 1″ × 2″ trim or molding at a local lumber store. You may even be able to have them cut it to the desired lengths.

To decide on the mold size, measure the inside dimensions of your kitchen sink, utility tub or whatever vat you plan to use. Be sure the measurements are accurate. To avoid any possible

Working on a flat surface, fit one of the short sections against a long section of trim. Nail together.

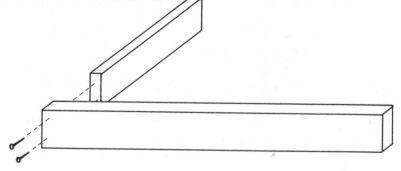

Set both mold sections on a flat surface and fit together to form a frame. Nail.

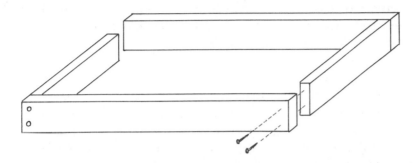

The completed Easily Constructed Mold. Note the watermark in the lower right-hand corner of the screen. Photo by Chris Miller.

errors, cut out a cardboard template to the measurements of the proposed mold and set this into the sink. There should be enough room around the edges for your hands to fit comfortably. Adjust the template if necessary.

Using the template as a guide, measure and cut the trim into four pieces of the proper lengths and sand the ends smooth.

Working on a flat surface, fit two of the lengths together—one of the short sections of trim against a long section, if you are making a rectangular mold. Nail these two pieces together. (See illustration.) Repeat with the other two sections of trim. Next set both L-shaped components together to form a frame. Nail in place.

Apply one coat of varnish or use a waterproof polymer coating to seal the wood. Let dry.

Roll out a section of fiberglass window screening and set the finished frame on top. Trace around the outside dimensions with chalk and cut out the screening. If you plan to add a watermark, do so before attaching the screen to the mold. Chapter 8 gives directions for making watermarks.

Set the screen on top of the frame and staple it to one side. Place the staples close together so that the screen does not sag. Pull the screen tight to the opposite side of the frame and staple in place. Repeat for the remaining two sides.

REINFORCED MOLD

If you plan to do a great deal of papermaking or if you want to make paper sheets by the pouring method, you will need a reinforced mold with either a deckle or a deckle box. Materials:

> 1½″ × ⅝″ trim—mahogany is preferred but pine can also be used (total length: 58″ or as required)
> ½″ dowels (total length: 40″ or as required)
> miter box
> clamps to hold joints while drilling, nailing
> saw

sandpaper
waterproof wood glue
hammer, nails
drill and ½″ dowel or auger bit, also a bit to match nail
    size
screening—either fiberglass mesh or brass (See Sup-
    pliers List in Appendix for brass screen sources.)
heavy-duty shears
60 brass upholstery nails (or as required)
varnish or polymer coating

Decide on the mold size. If it is to be different from the
dimensions given here, measure the sink or tub you plan to use
as instructed under Easily Constructed Mold.

Cut the trim into two 15¼″ lengths and two 13½″ lengths or
to the dimensions required. Miter the ends of each length and
then sand smooth.

Measure and cut three lengths of dowel, each 13″ long, and
sand the ends smooth. The dowels will be set into the longer
sections of trim, ½″ from the bottom edge. Measure for the
placement of the dowels as shown in the illustration and mark
the placements on the trim with a pencil.

If you are making a mold with different dimensions from
those given, plan to set in dowels at approximately 4½″
intervals along the longer sections of trim. If the mold is much
larger than the one shown here, you will need more than three
lengths of dowel.

Hold the two longer trim sections together to check the
marks for the dowel holes. Be sure that the marks line up on
the two sections.

Using the ½″ dowel bit and drill, make holes for the dowels
in both long sections of trim at the places marked. Drill halfway
into the trim. The tip of the bit will come through the other
side of the trim slightly. Sand the holes and fit in the dowels.
Assemble mold with the dowels in place and check the fit.
Make any necessary changes.

Remove the dowels and apply glue to both ends of each one. Reset the dowels into the holes and hold or clamp the sides of the mold firmly together while the glue sets.

Working on a smooth surface padded with a towel or newspapers, add the end sections of the mold to the sides to form a frame. Drill nail holes at each corner as shown. (See illustration.) Then nail the ends in place. Drill a nail hole into each dowel too. The holes formed by the tip of the drill bit on the outside of the mold when you drilled the dowel holes will guide you. Nail into each dowel.

Waterproof the mold with a coating of varnish or polymer. Let dry.

Diagram for Reinforced Mold.

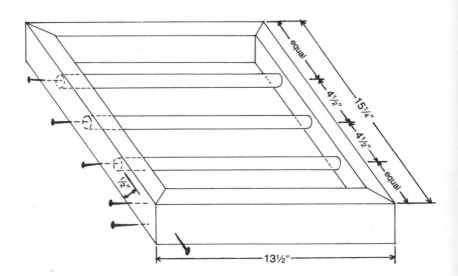

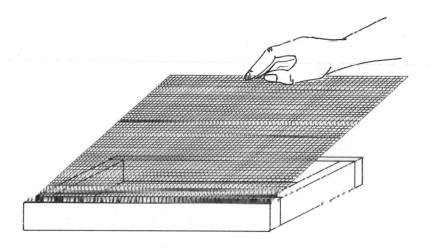

Pull the screen tight and staple to the opposite edge of the mold.

Cut out a section of screening 20″ × 17¾″. If you are using different mold dimensions from those given here, cut the screening about 2¼″ larger than the mold on each of the four sides.

Spread out the screen and center the mold on top of it. The dowels should be closer to the edge of the mold that is now uppermost (later to be the bottom of the mold). Turn up one edge of the screening. With your thumb, press an upholstery nail through the screen into the wood. This will hold the nail in

51

place long enough for you to hammer it in. Tack down one edge of the screen with nails placed about 1" apart.

Pull the screen as tightly as possible to the opposite side of the mold. Holding the screen tight with one hand, push in a nail. Hammer nail in, then tack down the remainder of this edge. Trim off any excess screening that extends over the inner edges of the mold.

Cut the screen as shown in Diagram A. Turn up the remaining edges and tack in place. Finish mold by trimming the screen as shown in Diagram B.

DECKLE BOX

A deckle is a thin frame that fits over the paper mold. This allows for the formation of thicker paper with a more uniform edge.

First tack two edges of the screen to the mold, then cut the screen as shown in Diagram A. Turn up remaining edges of screen, tack to the mold and trim as shown in Diagram B.

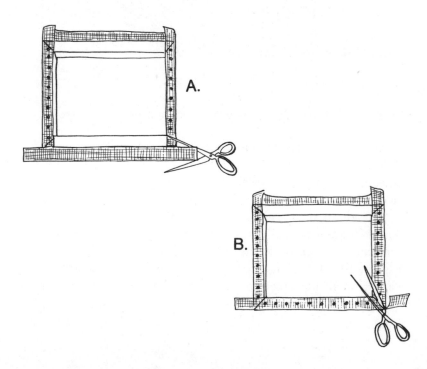

A.

B.

A deckle box is a deep container that fits over the mold. Pulp water is poured into this so that the deckle box keeps the pulp from mixing with the other water in the vat. The deckle box method gives more control over sheet formation and is preferred by some.

To make paper by the pouring method (described in Chapter 4) you will need a deckle box to use with the mold. Materials:

4½" × ⅝" wood (total length: 62" or as required)
1½" × ¼" trim (total length: 60" or as required)
hammer, nails
T-square
drill, and a bit to match nail size
clamps
sandpaper
varnish or polymer coating

The finished Reinforced Mold, screen-side down. Photo by Chris Miller.

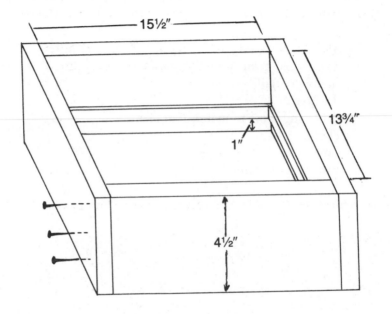

Diagram for constructing the Deckle Box.

To make a deckle box to fit the Reinforced Mold dimensions given here, cut the wood into four lengths (as shown); sand the edges, then fit these four lengths together to form a frame. (See illustration.) Drill nail holes as indicated, then nail the sections together.

Cut the trim into lengths (as shown) and fit the trim inside the deckle box 1" from the bottom edge. (See illustration.) Nail in place with small finishing nails.

The Deckle Box, set in place on top of the Reinforced Mold. Photo by Chris Miller.

Varnish the deckle box or coat it with polymer.

Before using the Reinforced Mold with the deckle box, set the mold on a flat surface such as the bottom of a sink or tub. Place the deckle box on top so that the trim rests on the top edge of the mold. The deckle box should fit snugly with no gaps between the two units. If the deckle box does not fit snugly, either make adjustments in the construction or attach strips of felt with waterproof glue to fill in any gaps.

THE DECKLE

Many papermakers using the vat method (see Chapter 4) prefer to use a deckle with their mold. The deckle is a framelike unit that fits on top of the mold and keeps pulp from flowing over the edges of the mold during dipping. For this reason, it is possible to form thicker sheets of paper by dipping when using a deckle. The deckle also creates a smoother edge on a paper sheet.

The deckle must fit snugly in place on top of the mold. You can build a deckle to fit the Canvas-Stretcher Mold, the Easily Constructed Mold or the Reinforced Mold. The dimensions given here are for a deckle that will fit the Reinforced Mold. Materials:

> 2" molding (Total length: 66")
> ¾" quarter-round (Total length: 66")
> hammer, nails
> T-square
> drill, and a bit to match nail size
> clamps
> miter box
> saw
> sandpaper
> varnish or polymer coating

Note: unless you have good-quality woodworking equipment and above-average woodworking skill, you may find it helpful to construct this deckle at a local frame-it-yourself store, if one is available.

Measure and cut four lengths of molding to match the dimensions of the Reinforced Mold (or of the mold you will use with the deckle). Miter the corners. Set the four sections together to form a frame and check the fit over your mold. Make any necessary adjustments, then lightly sand the edges of the molding.

Clamp two pieces of the molding together at right angles and drill for nail holes as shown in the illustration. Nail the pieces together. Repeat for the other two lengths of molding. Fit the frame together. Drill nail holes and nail the two remaining corners.

Cut four lengths of quarter-round to the same dimensions as the molding. Set the quarter-round against the flat side of the molding, flush with the outside edge. (See illustration.) Nail in place using small finishing nails. Check the fit of the deckle on the mold after you have added two pieces of quarter-round. Make any necessary adjustments. Attach the two remaining pieces of quarter-round.

Diagram for constructing the Deckle.

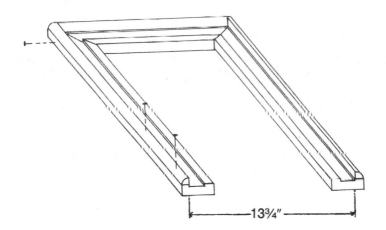

The Deckle, set topside down to show the attached sections of quarter-round. Photo by Chris Miller.

Varnish the deckle or use polymer to waterproof it.

See Chapter 4, under "The Vat Method," for instructions for use of the deckle.

BRASS VS. FIBERGLASS SCREENING

Fine brass mesh is used by papermakers who make their living at their craft because it produces a much finer-textured sheet of paper than is possible using fiberglass-mesh screening. (See the list of suppliers in Appendix for sources of brass screening.)

Another problem with using fiberglass screening is that with repeated use, the screening can stretch or sag toward the center of the mold. This can be a serious problem if you are using a deckle with the mold because the sagging screen does not make contact with the deckle edges and the pulp can, therefore, slip underneath, creating an uneven paper edge.

You can prevent this problem by using brass screening or by periodically removing the fiberglass mesh and either replacing it with a new piece or repositioning it, stretching the screen tight again.

4

Making Paper

There is no single correct way to make paper. You can use recycled junk mail to turn out attractive sheets of stationery or you can use 100 percent rag pulp with a neutral pH for museum-quality paper with a high degree of permanence.

And there is really no definitive paper "flaw." In fact, it is hard to fail at this craft at all, for what one craftsperson considers a failure—perhaps a torn sheet—another would be happy to have as just the right ingredient for a collage. Or yet another papermaker might use the torn sheet as the perfect dramatic background for a three-dimensional paper artwork. Even if you are not making collages or paper sculptures, though, you can still make use of your mistakes.

Torn sheets can be used for note cards or gift enclosures. Sheets with texture like a relief map can be used as watercolor paper. Paper with spots from bits of pulp that were not blended long enough can be called "Speckled Supreme" and used as interesting stationery.

In fact, paper made years ago is often flawed: stained by the drying ropes, rippled because it was not thoroughly dried before printing and spotted where water drops fell from the vatman's hands.

The flexibility of this craft is one of its great charms. Even beginners can turn out handmade paper that is unique and attractive. You really are a success no matter what you create on the paper mold. With this in mind, relax and enjoy the serendipity of papermaking—you may find that your "mistakes" turn out to be the most beautiful sheets of paper of all. And besides, if the paper were perfect, it would look as if it had been made on a machine.

PREPARING THE PULP

Paper Sources of Pulp. After you have accumulated a good supply of junk mail—envelopes, old letters, postcards, advertising circulars and so on—you are ready to make paper. Tear the junk mail and wastepaper into small pieces no larger than about 1" × 1". The smaller you tear the paper now, the easier it will be on your blender later.

Tearing a large quantity of paper is hard on your hands, so you may want to invest in an inexpensive paper cutter. Stationery and office supply stores carry small models (11" × 13") that are well worth the cost.

To prepare wastepaper and junk mail on a paper cutter, first feed several sheets of paper or envelopes under the blade and cut them into narrow strips. Then place ten to twenty of these strips in the paper cutter and cut into small squares about 1" × 1". You can feed the strips through in a very short time. If you work on a table or counter top, you can sweep the squares of paper into the boiling kettle by holding the kettle under the edge of the table or counter. Store any extra cut up paper in plastic bags for later use.

While you are preparing the paper for pulp, you can be selective about color and quality. If you want finished sheets that are bright white, then choose junk mail and wastepaper with little or no printing. Tear around illustrations and use a large quantity of unprinted paper and/or envelopes. Use only white paper and remove any postage stamps.

For an antique ivory-colored paper, add up to one-third colored junk mail to the batch of pulp. Look for subdued colors like pale yellow, light green and buff.

Junk mail in bright colors can also be used to make paper, and sheets from this pulp will be pastel shades of the original colors. Colored pulp can be used alone or as an addition to other batches for a variety of effects. Sort the mail by color. Choose red, green, orange and dark blue paper and keep the colors separate. Use a paper cutter or tear the colored paper into 1" × 1" pieces, then store each color in a plastic bag until you have enough for a batch of pulp. (Chapter 7 gives more information on making colored paper in a variety of ways.)

For paper with a gray color, use junk mail with printing and illustrations. Magazine covers and advertising circulars are good choices. Paper made entirely of this type of pulp will be light gray, often with a bluish tint.

While you are tearing or cutting the junk mail into small pieces, be sure to look for any staples, paper clips, plastic tape or cellophane windows. Remove these as they will not break down into pulp.

Eight to ten loosely packed cups of torn paper will make more than twenty new sheets. (This could vary, depending on the thickness of each sheet and the size of your mold.)

Other Sources of Pulp. All fabrics—wool, cotton or synthetic—that you plan to add to your recycled-paper pulp must be cut into *very small* pieces. Otherwise, the fabric fibers will overload your blender, possibly damaging the machine. (It was not designed to double as a Hollander beater.) So, cut fabric into squares no larger than ¼" × ¼". (To make rag paper, you need a beater. Chapter 10 describes methods of processing fabric on a larger scale.)

The dyes used to color fabric do not usually break down during the boiling stage of pulp preparation so if you want white pulp, use only white or light-colored fabric in that batch. However, small pieces of colored fabric will add speckles and texture to sheets of white paper, which could be a desirable effect.

63

String, wool or acrylic yarns and mohair should also be cut into ¼" pieces. Since many yarns and strings are more than one-ply thick, you can also separate the strands.

Lint from the clothes dryer should be examined carefully for any foreign objects. Remove anything that will not break down into small fibers.

BOILING THE PULP

After the fabric and paper to be recycled have been sorted and torn or cut up, they are boiled to break down the fibers, to remove some of the dye and to wash away any grease or dirt.

Place the squares of paper, bits of fabric, pieces of string or yarn in a large enamel or stainless steel pot. Do not use aluminum as it could become corroded.

Note: Lint does not need to be boiled.

Add water to cover the paper, plus 2" more. Push the paper down into the water with a stainless steel or wooden spoon so that it is evenly wetted. Then add 2 tablespoons of household bleach for every 8 to 10 cups of loosely packed paper squares. Do not add bleach to a batch of brightly colored paper if you want to make colored sheets.

There may be some odor while the pulp is boiling, but this is slight if present at all. Turn on a vent fan, if you have one.

During boiling, the water may turn color, from tan to red to dark brown, depending on the color and finish of the recycled paper in the batch. If the water does turn a dark color after the pulp has been boiling for 45 minutes to an hour, pour the pulp into a plastic or enamel colander set over the drain in the sink. Add fresh water to the kettle and return the pulp. Set the kettle back on the stove and continue boiling for the remaining time.

Let pulp boil gently for 1 to 2 hours (white paper will not need to boil as long as heavily printed or colored paper). Paper with printing will not be bleached completely clean during boiling. Printing will still be visible when the pulp is ready to use. However, pulp with visible printing can be used to make

Place the torn pieces of paper in a large kettle and add water to cover.
Photo by Chris Miller.

paper, for when the pulp is blended, the printing disintegrates as the fibers are broken down. If there is a great deal of printing in a batch of pulp, though, the resulting new sheets of paper will be less bright or even slightly gray in color. You can compensate for this by adding whiteners (see Chapter 6).

After the paper has been boiled, set a colander in the sink and pour the pulp and water into it. Let drain a few minutes, then rinse the pulp with cold, clean water. A spray attachment is helpful for this. When the pulp is cool enough to handle, squeeze it by hand to drain out all the discolored water. Then rinse again.

When the water from the pulp runs clear, form the pulp into several balls, squeezing out as much water as possible. Store these pulp balls in a sealed plastic bag in the refrigerator. Pulp will keep several weeks without spoiling.

SETTING UP THE WORK AREA

After the pulp is boiled and bleached, you are ready to start making paper.

To one side of the kitchen sink (or vat or tub), set out a pile of folded newspapers to act as a blotter, or use an old towel, folded. Next to this, stack the couching cloths and felts.

If you have a double sink, set a wooden cutting board over one half. This makes a convenient platform for draining the newly formed paper while it is still on the mold.

Set up your blender, get the pulp ready and assemble any extra materials you plan to use (Chapter 7 describes special additions).

BLENDING

Take a portion of boiled pulp from its plastic storage bag. About 1 tablespoon of pulp will make a sheet of paper approximately 8½″ × 11″. Place 2 to 3 cups of warm water in the blender. Do not fill the appliance completely as some blenders overflow during processing. Add the boiled pulp. Add any sizing, whiting, fillers or coloring at this stage. (See Chapters 6 and 7 for directions.)

Set the blender on "Liquefy" or the highest setting and blend for 15 to 30 seconds. The pulp mixture is ready to use when it looks cloudy and no individual pieces of paper or large fibers are visible. If you see pieces of pulp, continue blending for another 15 seconds.

Lint from the clothes dryer can be added to the mixture before blending, as well as small pieces of cotton or other fabrics or shredded bits of yarn. Just be sure that the blender motor is not overloaded. Watch for any signs of overheating—a burning smell or whining noise. If any of these overload symptoms occur, cut back on the ratio of pulp to water in your mixture or blend in shorter spurts.

Remember that large pieces of cotton, synthetics, wool or any other fabric will not process in the blender and will only ruin the machine.

After the pulp is well blended, you can add any extra ingredients that you do not want broken down. Some paper-makers add flower petals, feathers or pieces of colored yarn or string to give their sheets variety. Try these or anything else that is not too bulky. (Chapter 7 gives more suggestions and tells how to add different items to your paper.) Stir any creative additions into the pulp in the blender with a spoon. You are now ready to form your first sheet of paper.

Blend about 1 tablespoon of boiled pulp at one time. Photo by Chris Miller.

With one hand hold the deckle box and mold on the bottom of the sink or tub while pouring the pulp with the other hand. Photo by Chris Miller.

FORMING THE PAPER

There are two basic ways to make paper: the vat method and the pouring method.

The Pouring Method
To use this method of papermaking, you will need the Reinforced Mold with the deckle box. Some craftspeople find it easier to form a good sheet of paper with even thickness using the pouring method.

To make a sheet of paper on a Reinforced Mold of the

dimensions given in Chapter 3, use ½ cup of firmly packed pulp. In a blender, blend at highest setting 1 tablespoon boiled pulp with 2 to 3 cups of water. Pour the resulting water-and-pulp mixture into a large pitcher. Repeat until the whole ½ cup of pulp has been blended. Now you are ready to form a sheet of paper.

Fill the sink, utility tub or bathtub with water. Set the mold and deckle box together, with the deckle box resting on the top edge of the mold. Set the entire unit on the bottom of the sink or tub. The water should come at least halfway up the sides of the deckle box—the mold must be completely covered with water.

To form the paper sheet, with one hand, hold the deckle box and mold on the bottom of the sink or tub and with the other hand, pour the entire pitcherful of pulp and water into the deckle box.

The pulp will spread out in the water contained in the deckle box. Usually, the best sheet formation occurs right away, so be alert. When the pulp seems evenly dispersed in the water, lift the mold and deckle box out of the tub, using both hands and keeping the unit level. There will be quite a pull due to suction as you bring the mold to the water's surface.

Remove the mold unit from the water and set aside to drain. Take off the deckle box and inspect the new sheet of paper for defects—hair, lint, tears, holes or thin areas. Hold the mold up to a light source and check for even sheet formation. The paper sheet will not fall off the screen during this inspection.

If you are satisfied with the paper sheet formed, set the mold on the counter and then couch the new paper as described in the next chapter.

If you are not happy with the paper, scrape the pulp off the mold screen with your hand or a blunt knife. Return the pulp to the blender in several batches and reblend for just a few seconds to mix the pulp and water. Depending on the type of pulp you are using, stirring with a spoon may be sufficient. Set up the mold and deckle box again, place the unit back in the sink or tub and re-form the sheet.

When the pulp spreads out evenly in the water, raise the deckle box and mold out of the tub or sink. Photo by the author.

Remove the mold from the deckle box to view the formed sheet of paper. Photo by the author.

When the pulp is evenly dispersed in the water of the sink, use both hands to lift the mold up out of the water, holding it level. Photo by Chris Miller.

Uneven pulp distribution results in sheets of paper with some areas thicker than others. Photo by Chris Miller.

If a good sheet of paper does not form immediately after you pour the pulp and water into the deckle box, gently stir the pulp water in the box with your hand to get an even distribution. Or you can lightly shake the entire mold in the water. Watch the pulp carefully and when it shows a uniform dispersion, lift the mold unit out of the water.

The Vat Method

Some people find it easier to get a good sheet of paper using the vat method with any of the molds in Chapter 3, with or without a deckle.

Put 2 to 3 cups of water in the blender. Add 1 tablespoon of pulp and blend at the highest setting for 15 to 30 seconds, until no large pieces of pulp are visible.

Set the stopper in the drain of a clean, grease-free sink and fill it half full with warm water. (It is more comfortable to work in warm water.) Add 3 to 5 blender-loads of the blended pulp-plus-water mixture.

Using any one of the molds in Chapter 3 with the screen side up, dip the mold into the pulp water in the sink. With one hand, hold the mold on the bottom of the sink. With your other hand, gently stir the pulp water above the mold until you see an even dispersion of pulp. (The mixture should look uniformly cloudy.)

When the pulp is evenly dispersed, use both hands to lift the mold up out of the water, holding it level. The suction caused by the rush of the water away from the mold pulls the pulp onto the screen to form a sheet of paper. If the paper formed is to your liking, remove the mold from the water and let it drain on the wooden cutting board over the other side of a double sink or set the mold to the side of a single sink.

After the paper has drained a few minutes, hold the mold with the sheet in place up to a light source. The paper will not slip or fall off. If the sheet is flawed in any way—if it has holes, thin areas, dirt, hair or rips in it—then turn the mold upside down and set the screen against the surface of the water in the

72

sink. The suction created by the water as you pull the screen away will remove the pulp. Before returning pulp to the sink, however, be sure to remove from it any hair, dirt or other foreign matter. Then redip the mold and form another sheet of paper.

To use the deckle: Set the deckle on top of the mold. Holding the two together as a unit, dip the mold and deckle into the vat of pulp and form a sheet of paper. Bring the unit out of the water and set to the side of the sink or tub to drain. Remove the deckle and check the paper formation.

Directions for removing a wet sheet of paper from the mold are given in the next chapter.

You can dip three to five sheets of paper in one sinkful of pulp and water before adding more blended pulp to the mixture. Or add more pulp when the sheets become too thin or tear during couching. To make thicker sheets of paper, just use greater pulp-to-water ratio. Do not, however, blend more pulp at any one time in the blender. Instead, put less water in the sink at the start.

After dipping and forming each sheet of paper, turn the mold upside down and rinse it over the sink to remove any bits of pulp that may cling to the screen. A spray attachment is helpful. Bits of pulp left on the screen could create flaws in the next sheet of paper formed.

After long use, molds used in the vat method can acquire a built-up rim of pulp under the screen where it is tacked or stapled to the wooden frame. Clean out this pulp occasionally with a toothpick or blunt knife. Be careful not to damage the screen.

Straining the pulp from the white water (water left after papermaking) will save your pipes from possible clogs. When you are finished making paper with the vat method, cut a piece of fiberglass-mesh screening about 2" larger on all sides than the drain in your sink, utility tub or bathtub. With the pulp water still in the sink, hold the screen over the drain and pull up the stopper, through the screen.

If the water slows down too much during draining, replace the stopper and remove the screen from the sink. Then remove the pulp that has collected on the screen. This pulp can be discarded or reused later. Replace the screen and continue draining the sink.

To reuse strained-out pulp, store in sealed plastic bags in the refrigerator. It will keep for several weeks. When you are ready to make paper again, put the preblended pulp into the blender with 2 to 3 cups of water. Set the blender on low and process the pulp just long enough to mix it with the water. Now it can be used with either of the methods.

EASY PULP

This method of pulp making is very easy and requires a minimum of time and equipment. It is, therefore, especially well suited to papermaking demonstrations or for use by a school class or scouting group.

For the pulp, use facial tissues, any color. Be sure, though, that they are not wet strength. You can also use toilet tissue, but this requires the use of a blender. Paper towels do not usually break down well and should be avoided.

Take ten sheets of facial tissue and add to a sink or pan full of water. Let soak for 10 minutes, then beat with an eggbeater or a hand electric mixer. (If you are using an electric mixer, be sure the area where you are working is dry. Do not use an electric mixer with a frayed cord, especially around the sink or pan filled with water.) Beat the pulp until it is completely broken down and no large pieces are visible.

Using any of the molds described in Chapter 3, dip and form a sheet of paper in the vat filled with tissue pulp. Or you can pour the pulp water into a mold with deckle box. The sheets of paper can be couched and dried as for paper made from recycled junk mail. If you use colored facial tissues, the paper will be the same color, but slightly lighter.

To use toilet tissue to make pulp, pull twenty sheets or so off a roll and add to a pan or sink filled with water. Let soak 15 minutes. Then place the pulp in a blender, ¼ cup at a time. Blend on "Liquefy" or the highest setting for 5 to 10 seconds or until no large pieces of pulp are visible. Pour the pulp back into the sink or pan and form paper.

Paper sheets made from tissue pulp are soft and flexible with a very pleasing surface texture. Paper made from toilet tissue pulp is especially soft to the touch.

Unsized sheets of paper made from tissue pulp can be used with ball-point pen. There may be some bleed-through to the back of the sheet with a hard-tip felt pen, but this depends on the sheet thickness. Lines made with soft-tip felt markers spread on the paper and bleed through to the back, but not badly. See Chapter 6 for ways to size these papers for less absorbency.

One of the main drawbacks to using tissues as a pulp source is that they are more expensive than recycled mail and wastepaper. However, the finished paper has such a delicate, pleasing texture that it may justify the extra cost. If you want to use tissue for pulp, look for sales on tissue products or buy store brands that are less expensive.

THICK AND THIN PAPER

You can easily adjust the thickness of the paper you create simply by changing the ratio of pulp to water in the vat or in the pitcher. However, *don't* blend more pulp at one time in the blender. Always process only about 1 tablespoon of boiled-paper pulp to 2 to 3 cups of water. To increase the ratio of pulp to water in either method just use a larger number of blenderfuls per sheet.

If you want to make thick paper, do not add water to the sink in the beginning. Fill it only with water-and-pulp mixture from the blender. Thick paper is especially well suited to use with watercolors or to make mats for framing pictures.

To make thin paper, simply add fewer blender-loads of pulp and water to the vat. Be careful, though, that you add enough pulp to completely cover the screen of the mold. If you have not added enough pulp, the sheet of paper formed will be incomplete—with thin areas and ragged edges that do not fill out the mold.

You can also make thin paper sheets by continuing to dip sheet after sheet in the same pulp-and-water mixture without replenishing the vat. The sheets will become progressively thinner.

PAPERMAKING PROBLEMS

Foreign Matter in the Blender. Papermakers have always had trouble with unwanted objects finding their way into the paper. One of the primary avenues for this adulteration was the beater. If any foreign material—an insect, leaf, twig or whatever—fell into the beater during operation, it was pulverized into thousands of small pieces that were impossible to retrieve. When rags were scarce, the pulp could not be discarded, so old paper often contains bits of "extras."

You have more control over the water and boiled paper you put into your blender. Check carefully to see that nothing unwanted goes in, for after blending has been completed, it will be even harder to remove foreign matter.

Hair. Hair can be mixed into the pulp when lint from the clothes dryer is used or it can float in from the air.

Strands of hair cannot be easily removed from a finished sheet of paper without damaging the sheet itself. Therefore, it is important to check for hair that may be embedded in the pulp at the time the paper sheet is formed. Make paper in a well-lighted area. After each sheet is formed, hold it up so that light shines through the paper. Hair and other unwanted matter will be plainly visible, as will any surface defects like thin areas.

When one or two stray pieces of hair float into the pulp as you form a sheet of paper, they do not pose a difficult problem. Just remove the hair and re-form the sheet of paper. However, sometimes bits of hair can contaminate an entire batch of pulp. To minimize this problem, do not tear or cut up paper in areas where people brush or trim their hair or where a pet could be shedding.

If you do make a batch of pulp and find that it contains large amounts of hair, you can still salvage the pulp for use. Just add enough of some other highly textured material to camouflage the hair. Try adding dried grass or straw, flowers with their stamens, leaves or bits of yarn and thread. See Chapters 7 and 11 for directions on how to add these items to pulp.

Lint and Threads. The most common source of lint and threads is the couching cloth you use to remove the sheet of paper from the mold. So, check these cloths before you start making paper and remove any loose threads. Cutting the edges with pinking shears may slow down the unraveling process somewhat, but depending on the type of fabric used, pinking does not altogether eliminate the problem. You can also stop unraveling by using a sewing machine to zigzag stitch the edges of the cloths or you can turn under the raw edges and machine stitch them in place. Make sure that a turned edge or zigzag stitching does not come in contact with wet sheets of paper since these could leave unwanted impressions.

Lint or threads picked up from couching cloths can usually be pulled off the surface of a dried sheet of paper quite easily since they are not formed into the paper. However, lint and pieces of thread may leave an impression on the paper, most visible when the sheet is held to the light. It is virtually a form of watermark and cannot be removed without destroying the sheet of paper.

SHEET FORMATION PROBLEMS

Problem	Cause	Cure
Pulp does not completely fill the mold.	Not enough pulp for mold size.	Add more pulp to the vat or pitcher.
Bubbles appear in formed sheet.	Air trapped under the pulp.	Gently shake the mold against the water surface to break up air bubbles.

Problem	Cause	Cure
Large pieces of fiber or paper in formed sheet.	Pulp not blended long enough.	Return pulp to blender. Blend another 15 seconds.
Some areas of the sheet thicker than others.	Uneven distribution of pulp.	Re-form sheet, stirring the pulp water to disperse pulp evenly.
Holes in paper sheet.	Pulp not evenly distributed.	Re-form sheet. Stir the pulp water thoroughly.
Hair in sheet (or other foreign matter).	Contaminated pulp.	Remove foreign matter. Re-form sheet.
Lint or string on sheet surface of dried paper.	Unraveling couching cloths.	Pull out string or lint from sheet surface. Trim couching cloths —pink edges or turn under and stitch—before couching more paper.

5

Couching and Drying
the Paper

No matter which mold or which method of papermaking you chose, Chapter 4 left you with a wet sheet of paper. This chapter will explain how to turn that wet sheet into a finished sheet of paper.

ON-MOLD DRYING

The simplest method of drying the new paper is to leave it on the mold screen. This method was used by ancient paper-makers and it is still used today in some Oriental countries. The big drawback to on-mold drying is that it ties up your mold. You can either make just one sheet of paper every 3 to 10 hours or longer or you must have several molds available. An alternative is to use a mold with a removable screen. The Embroidery Hoop Mold is ideal for this purpose and is recommended for beginning papermakers and for children.

With the Embroidery Hoop Mold, you can dip and form a sheet of paper, take the hoops apart and remove the screen with the wet paper intact. You then place the paper in a sunny spot to dry. Depending on the weather, this drying period can take 10 hours to several days. But with another screen inserted into your mold, you can form another sheet of paper immediately.

COUCHING

When early papermakers learned to remove or couch a sheet of paper from the mold, their productivity increased. New sheets of paper could then be formed continuously on the same mold and the couched sheets could be stacked, pressed and dried.

For a simplified version of the traditional couching method, you will need several felts and couching cloths, all at least 1″ larger on all sides than the sheet of paper you are forming. You will also need several old towels or some extra pieces of felt.

Spread a couching cloth over the wet sheet of paper and then gently press in place. Photo by Chris Miller.

On a flat surface next to the sink or tub, lay a towel or a pile of newspapers to absorb excess water. Place one of the felts on top. After you have made a sheet of paper using either the vat or pouring method, remove the deckle or deckle box if you are using one, then set the mold over the sink to drain.

Spread a couching cloth over the wet sheet of paper and press gently in place until the cloth is completely damp. Next lay one section of a towel or a piece of doubled felt, several paper towels or even a sponge on top of the couching cloth. Press lightly

After blotting the couching cloth with a towel or piece of felt, carefully pull the couching cloth away from the paper mold. The paper sheet should adhere to the cloth. Photo by the author.

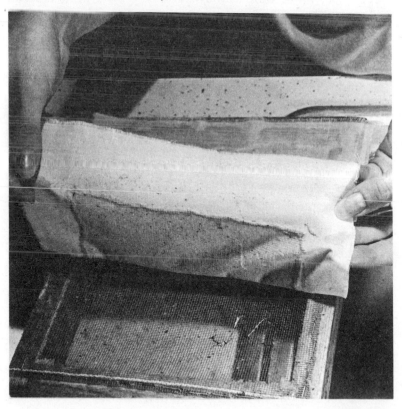

until water is absorbed and the blotting material is damp. Remove the felt, towel or whatever and repeat process with dry blotting material. The blotter absorbs water from the paper through the couching cloth with the result that the paper sticks to the couching cloth and is easy to remove from the screen.

Set aside the towel, felt or sponge, then lift up one corner of the couching cloth. If the paper sheet is adhering to the cloth, gently pull the couching cloth with the paper attached away from the mold. Place the cloth with the paper side down on top of the felt that you laid next to the sink.

If the paper sheet is not adhering to the couching cloth when you lift one corner, then carefully scrape up a corner of the paper with a blunt knife or your finger to get the sheet started.

To make couching easier, use a blunt table knife and carefully push the paper sheet in from the edges of the mold about ¼".

Thin paper sheets are more difficult to couch than thick sheets. If the paper tears while you are couching it, you probably need to add more pulp to the water. If the sheet rips too badly to be salvaged, scrape the pulp from the mold screen for reblending. Rinse the screen and form another sheet of paper.

A uniform sheet of paper will couch more easily than one with thick and thin areas. It is important to check the sheet for flaws after you have formed it. Hold the mold to the light to check for thin areas, as these trouble spots tear easily during couching. In some cases, it might be less trouble just to re-form a sheet rather than risk couching problems.

Couching cloths and felts should be used only once then washed and, if necessary, pressed. Check your couching cloths and felts for lint or stray threads each time before you use them, and remove these and any foreign matter that could stick to wet sheets of paper and create flaws.

As you couch sheets of paper, stack them between couching cloths and felts, creating a pile. The wet sheets can be kept stacked for several hours before you dry them.

Couching the Embroidery Hoop Mold

The sheet of paper formed on a round embroidery hoop is sometimes difficult to couch until you reach the halfway point. The edges of the sheet have a tendency to stick to the mold screen, tearing the paper. If you have this problem, use your fingers to lift the paper sheet off the mold as you remove the couching cloth. A little help in getting the paper started is usually sufficient.

DRYING

There are several different ways to dry newly formed sheets of paper. Iron drying is fast (about 5 minutes a sheet) but involves more work. Oven drying is slower, but you are free to go do other things while the paper dries in 1 to 2 hours. And air drying is slower still, but it uses free energy.

There is little difference in the appearance of paper dried with an iron, in the oven or by the air, so choose the method that is most convenient for you.

Air Drying

The oldest method of drying paper, air drying is not recommended during humid weather. Under these conditions the paper takes so long to dry that mold has a chance to grow. However, if the weather is dry and sunny, you can easily air dry your paper with good results.

Choose a dry, dust-free spot with good air circulation. Form the paper sheets and couch them from the mold. Lay the paper out on a table or counter, couching cloth side down with the paper exposed to the air. Weight down the edges of the couching cloths with rocks, spoons or anything heavy enough to keep the paper from curling as it dries.

On a breezy day, paper will dry in a few hours. Check in 30 minutes to 1 hour. If the paper sheet lifts easily from its couching cloth, turn the sheet over to let the other side dry. Remove the sheets when they are completely dry. Place between pieces of typing paper and set a book on top to keep the sheets from curling.

Iron Drying

For this method you will need an iron and a wooden cutting board.

Set the cutting board on a counter, tabletop or ironing board. Be sure to iron paper in a dry location. Place a dry couching cloth on top of the cutting board. Then take a newly formed sheet of paper, still attached to its couching cloth, and place it paper side down on top of the cloth you have laid on the board. (See illustration.)

With the iron set on "Cotton," using no steam, begin pressing the back of the wet couching cloth. Keep the iron moving at all times to avoid scorching the paper.

Press a piece of wet paper between two couching cloths.

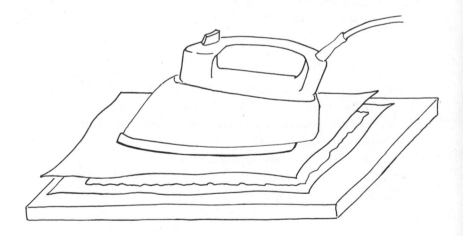

After you have ironed the top couching cloth for a minute or so and quite a bit of water has evaporated, flip the stack so that the bottom couching cloth is now on top. Iron the back of this cloth too.

Continue flipping and ironing until the paper sheet is dry. Test this by pulling up one corner of a couching cloth. If the paper comes away easily, it is dry. Make sure that there are no small wet spots remaining, though. These appear as darker areas than the surrounding sheet.

It takes about 5 minutes to completely dry a sheet of wet paper.

When the paper sheet is dry, peel away the couching cloths and place the sheet between two pieces of typing paper. Place under a heavy weight, such as a book, to keep the new paper from curling or wrinkling.

Oven Drying

This method takes longer but with much less work involved.

To oven dry, you will need a baking sheet and several large (2") paper clips (available at office supply stores).

Couch several sheets of new paper and stack these with felts. When you have finished papermaking for the day, transfer the stack of paper, couching cloths and felts to the clean baking sheet. Using the large paper clips, clip the edges of the couching cloths together in several places. (See illustration.) This will put enough pressure on the paper to keep the sheets from curling as they dry.

Set the oven at 250° F. and place the baking sheet on a middle rack. If you are drying a stack of twelve or more paper sheets, set a timer for about 1 hour. At that time, check the top one or two sheets to see if they are dry yet.

After about 1 hour, the top sheets of paper should be dry and these can be removed, together with the alternate felts and couching cloths. Leave a felt or couching cloth on top of whatever wet sheets are to remain in the oven.

You don't need to protect your hands to remove the cloths and felts since the fabric is not hot. Pull the cloths gently away

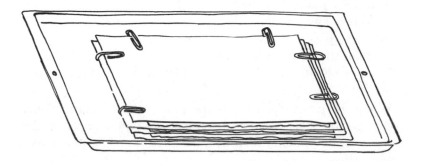

To oven dry the paper, stack each sheet between a felt and couching cloth. Set on a clean baking sheet and clip the edges together with large paper clips.

from the paper clips. Be very careful, however, not to touch the clips since these are quite hot and could burn you.

Flip the stack of paper sheets over and check the bottom one or two sheets. These will probably be dry too. Remove these sheets in the manner described, then return the baking sheet to the oven. Check the paper again in another 15 minutes or so. Continue removing dry sheets of paper until the entire stack is dried. The drying time will decrease rapidly as the stack gets smaller.

After you have removed a dry sheet of paper and its couching cloth and felt from the oven, carefully pull the cloth away from the paper. Place each new sheet between two pieces of typing paper and weight down the stack with a heavy book.

Total drying time for a stack of about twenty sheets of wet paper is 2 to 3 hours. Small batches of two to five sheets of paper dry in about 30 minutes.

Paper formed on the Embroidery Hoop Mold can be dried on the screen, either in the air or in the oven. To oven dry, form a sheet of paper then carefully take the embroidery hoops apart and remove the screen. Set the screen with its paper on a baking sheet. Place in a 250° F. oven. The paper will be dry in 20 to 25 minutes. To remove the sheet, carefully pull back the screen and peel off the paper.

REDRYING

If a sheet of paper dries with wrinkles or creases or if it curls, it can be redried to correct these faults.

To redry paper you will need a can of aerosol spray starch and a stack of newspapers.

Spread out the newspapers and lay the paper sheet to be redried in the middle. Thoroughly soak the sheet of paper with the spray starch. Turn the paper and spray the other side too.

Place the paper between two couching cloths and dry again using either the oven or iron drying methods. (Air drying could

This paper sheet developed wrinkles because it well not pressed long enough after drying. It is a good candidate for redrying. Photo by Chris Miller.

duplicate wrinkles or creases since not as much pressure is applied to the damp sheet during this method.)

PAPER AND HUMIDITY

Cellulose fibers are hygroscopic. This means that they readily absorb water from the air—the amount absorbed depending on the relative humidity. Because of this characteristic, finished sheets of paper vary in water content from day to day, changing with the weather.

These changes in water content can cause swelling and shrinkage of the fibers, resulting in puckered or curled sheets of paper. If paper develops these flaws, redry it using spray starch.

PAPER STORAGE

After you have spent so much time creating beautiful pieces of paper, you will probably want them to last awhile. Here are some tips on the care and storage of paper that will help promote maximum shelf life.

Make sure hands are clean when working with paper. Grease, oil or dirt from your hands could damage the sheets.

Never store batches of colored paper together with other paper, or white with colored. Some dyes are not permanent and they can easily bleed onto other sheets. Separate individual lots of colored paper from other paper with one or two sheets of white typing paper. Or better yet, store each lot in a separate folder.

Do not store paper in contact with paper clips, staples or rubber bands since any of these items could stain the sheets. Clear tape, gummed and masking tape can also leave stains.

Light is responsible for the fading of colored paper. Natural dyes are especially susceptible to fading. The ultraviolet rays in sunlight can also cause deterioration of the paper itself, so store paper in a dark location.

If you are using colored, handmade paper for artwork, be

careful where you display your art. Do not hang pictures on walls opposite windows, since the light there will be greatest and fading of colors most severe. To minimize fading, consider rotating your artwork, displaying a selection for no more than a few months, then storing it and displaying another work.

Beware, too, of fluorescent lights. These, like the sun, are a source of ultraviolet light. Cover fluorescent lighting fixtures with plastic sleeves which act as filters. Or substitute Plexiglass sheets (which filter out ultraviolet light) for the glass used in picture frames.

Always choose a dry location for storing paper. Choose wooden drawers and cabinets as storage areas since moisture can easily condense on the inside of metal drawers. The best temperature at which to keep paper is 68° to 72° F. (22° C.) with 50 percent to 55 percent relative humidity.

Excessive humidity is another enemy of paper. Besides causing it to swell, high humidity (above 70 percent) can also allow mold to form. Air conditioners or dehumidifiers will help keep paper dry in damp climates or during summer months. Good air circulation also inhibits the growth of mold.

Mold shows up first as dull, rusty spots that discolor the paper sheet. This is known as "foxing." The mold actually feeds on the sizing and paper fibers, thereby weakening the sheet.

To treat paper that has been attacked by mold, first move the paper to a dry place. Allow air to circulate freely all around by setting the paper on a rack or screen. Expose the paper to direct sunlight for about 1 hour or place the paper in a closed container for 2 to 3 days with some crystals of thymol. Thymol is a fungicide, but since it is volatile, it does not offer permanent protection. Therefore, do not return the paper to a humid location or the mold is likely to grow again.

Since dust contains the airborne mold spores, good housekeeping is an effective preventive measure for keeping stored paper mold free.

Do not expose handmade paper or artwork to high temperatures, since this will also accelerate the deterioration of the paper. Avoid storing or displaying paper near radiators, heating registers, fireplaces or air ducts.

Pollution is yet another cause of paper deterioration, and sulfur dioxide, a gas produced by the combustion of oil and coal, is the most damaging pollutant in the atmosphere. Sulfur dioxide is especially prevalent in cities where smog is a problem. Sulfur dioxide causes discoloration of paper, makes sheets brittle and eventually causes the disintegration of the paper fibers. Even after paper is removed from contact with sulfur dioxide, the pollutant still can cause damage since the gas is absorbed by paper and is converted into sulfuric acid, which does not evaporate.

Rare books and important documents are stored in rooms where the air is filtered to remove sulfur dioxide. Even if you can't live in the country where the air is cleaner, you can still use air conditioning to reduce the chance of damage to paper by sulfur dioxide.

Insects can also harm paper—the most common culprits being cockroaches, silverfish, termites and woodworms.

Roaches like damp, dark and warm places, so storing paper in a dry spot will discourage infestation. Roaches cause surface damage to paper and artwork on paper. If cockroaches are a problem in your locality, regularly spray storage areas with insecticides.

Silverfish also prefer warm, dark and damp locations. These insects cause damage to books and paper by eating the paper to get the sizing, though they also like the taste of bleached wood pulp itself. Store paper in a dry spot, off the floor. Check frequently for signs of silverfish and if they are evident, use a spray insecticide recommended to control these insects.

Termites and woodworms like paper because it, like wood, is composed of cellulose. However, they are more likely to attack books and framed pictures than stored paper. Good ventilation, especially in basement areas, is an important deterrent.

COUCHING AND DRYING PROBLEMS

Problem	Cause	Cure
Paper sheet peels off in sections during couching.	Not enough pulp; sheet is too thin.	Add more pulp to the vat or pitcher. Re-form sheet.
Holes in sheet.	Thin areas in paper torn during couching.	Check paper for thin areas before couching. If not of uniform thickness, re-form sheet.
Ripples in paper sheet after it has dried.	Humidity caused paper to change shape or ripples were pressed in during drying.	Redry sheet.
Round spots, visible when paper is held to the light.	Water drops from paper-maker's hands left marks in damp paper.	Dry hands before couching.
Rust colored stains on paper.	Foxing caused by mold.	Expose paper to sunlight and good air circulation or treat with thymol.

6

Sizing the Paper

WATER LEAF

The paper you made in Chapter 4 did not have any sizing and is known as water leaf. It is very absorbent, almost like a blotter, and often has a soft, limp feel to it. Water leaf is fine for many purposes and some hobbyist papermakers produce only water leaf.

Water leaf is preferred for printing. In old paper mills, water leaf was sent on its way to the printer's without further treatment, often without even being completely dried. Ball-point pen works very well on this paper. Hard-point, roller-type felt-tip markers work fine on water leaf too, with little or no bleed-through to the back of the sheet. Soft-tip felt markers do bleed through, though, as do the wide-tip felt pens. In fact, the wide-tip markers bleed through the paper so much that the design is almost the same on both sides. This may not be a problem, depending on your project. Water leaf would not be a good choice for stationery if you plan to use soft-tip felt markers, but it could be used successfully as paper for a felt-tip painting if the work were to be framed later.

93

The bleed-through problem is the same for watercolors and acrylic paints used on water leaf paper, although if the acrylics are applied undiluted, there is little difficulty. Some watercolorists prefer an absorbent paper for their work and water leaf does respond well to the paints. So it is a matter of preference on the part of the artist. Tempera poster paints also bleed through to the back of the paper sheet. Since the paper is so absorbent, pen-and-ink does not work well at all on water leaf, with large ink spots resulting.

Water leaf paper is a fine surface for drawing, though, and pencil, charcoal (soft, medium and hard), pastels, colored pencil and even glass-marking pencils can all be used with great success.

PAPER SURFACES

Paper formed on a mold with a fiberglass-mesh screen is distinctly different, one side of the sheet from the other. The side of the paper formed next to the screen takes on the imprint of the mesh—a small, bumpy pattern of raised and depressed areas that feels rough. This side is said to have "tooth." It is an excellent surface for charcoal and pastels.

The side of the paper formed away from the screen is much smoother, with no surface pattern. This side does not carry the imprint of the mesh, even in very thin sheets of paper. Pen-and-ink can often be used more successfully on the smooth side of the paper. Depending on the preference of the artist, the smooth side of the paper may also present a better surface for pencil drawing.

SIZED PAPER

By adding the appropriate sizing, you can use handmade paper with all media—even oil paints. The Romans used a flour-and-vinegar mixture to size papyrus. In the Middle Ages, papermakers used a glue made from hide trimmings as a sizing. In fact, before 1800 all paper was sized either with animal glue or a vegetable-gum solution. In 1800 a German, Moritz Friedrich Illig, discovered that rosin and alum could be used to size paper. However, it was another 25 years before Illig's method for sizing became widely used.

Today a typical commercial paper mill uses a sizing solution consisting of a rosin soap dispersion mixed with the pulp in a one-to-five ratio. The rosin soap has no natural affinity for the pulp fibers so a coupling agent must be used to attach the sizing to the fibers. This agent is usually alum (aluminum sulfate).

Basically, there are two methods of sizing paper: internal and external. Internal sizing is done at the time the pulp is blended. It is easy and does not require a separate step or extra drying, so many papermakers prefer this method.

Internal Sizing

There are many household substances that you can use to size paper: unflavored gelatin, cornstarch, white glue, animal glue and varnish.

Gelatin and Starch Sizing. This easy-to-make recipe can be used for both internal and external sizing. Paper sized with the gelatin-and-starch mixture has a crisper feel to it than water leaf, yet the paper is not stiff to use.

Sheets sized internally with gelatin and starch have a nice feel—firm yet flexible. They take ball-point pen well and also pen-and-ink. There is little bleed-through with hard-tip felt markers, but quite a bit with the soft tip. Watercolors bleed through badly, so does tempera. Acrylics can be used if they are mixed with a minimum of water. Pastel and pencil as well as charcoal are all suitable media for this paper.

Gelatin and Starch Sizing

1 packet unflavored gelatin
2 tablespoons cold water
2 tablespoons boiling water
2 tablespoons cornstarch
1 cup boiling water

Pour the packet of unflavored gelatin into a heat-proof bowl or a small saucepan. Add 2 tablespoons of cold water and stir until the gelatin forms a paste. Then add the 2 tablespoons of boiling water, stirring until the gelatn is dissolved. Stir in the cornstarch until it is also dissolved. Then add 1 cup of boiling water, mixing well. Be sure all the lumps are dissolved.

The sizing is now ready to use. For the vat method of papermaking, add 1 tablespoon of sizing to each blender-load of pulp and water. For the pouring method, use 1 tablespoon of sizing for each sheet (for the Reinforced Mold in the size given). Adjust this amount if your mold is larger or smaller or if you are making thicker or thinner sheets of paper (see Chapter 4). You can experiment with different amounts of sizing per sheet to see what the effects are.

The gelatin-and-starch sizing becomes solid as it cools. If you store the sizing in a covered container or in a small saucepan, it will keep in the refrigerator for several weeks without spoiling. To reuse, simply heat the sizing on a stove burner at a low setting for 2 to 3 minutes until the sizing is liquid again. Remove from the heat and use as instructed.

Variations. Mix 1 package of unflavored gelatin with 1 cup of boiling water. Stir until the gelatin is completely dissolved. Use 1 tablespoon of this sizing per sheet of paper.

Use cornstarch alone: add 2 teaspoons of cornstarch to 1 cup of hot water. Mix thoroughly. Use 1 tablespoon of the sizing per sheet.

Add a few drops of animal glue to the cornstarch sizing for more waterproofing. Or add white glue to the gelatin and starch sizing. Add 2 to 3 drops of glue to the finished sizing and stir well.

White glue alone can be added to a vat of pulp. Use 2 to 3 drops of white glue for each 10 to 12 cups of pulp water. Stir well.

Linseed Oil Sizing. Use ½ teaspoon of linseed oil to one blender-load of pulp. The linseed oil coats your hands slightly as you work with the paper and the oil also has a rather strong odor that lingers even after the paper is dried. Add perfume to the sheets if this odor is objectionable. Paper sized with linseed oil has a very soft feel and is sometimes quite limp.

THE COMPLETE BOOK OF HANDCRAFTED PAPER

You can use ball-point pen, pencil, pastels, charcoal and hard-tip felt markers on paper sized this way. Soft-tip felt markers bleed through as do watercolors, tempera and pen-and-ink. Acrylics may be used if not diluted with water.

Adding cornstarch to the linseed oil reduces the limpness of the sheet of paper while retaining most of the soft texture. In addition, the cornstarch makes the paper more waterproof so that soft-tip felt markers as well as pen-and-ink can be used with no bleed-through. Watercolor can also be used on paper sized this way. Add 2 teaspoons of cornstarch plus ½ teaspoon of linseed oil to one blender-load of pulp to be added to the sink.

For the pouring method, use ¼ teaspoon of linseed oil with or without 1 teaspoon of cornstarch for each sheet.

External Sizing

Another way you can size paper is to use a brush to coat the paper sheet after it is dry. With this method, one side of the paper can be coated (C1S—coated one side) or both sides can be coated (C2S).

Paper that is coated on one side only can develop a curl when it dries because the sizing has upset the surface tension of the sheet. You can prevent this by applying sizing to both sides of the paper.

Any of the internal sizings already described can be used to coat sheets of paper that have already dried. Mix up the sizing according to directions and instead of adding the sizing to the vat or blender, just pour it into a small container.

Use a fairly wide, soft brush. For small sheets (about 6″ × 9″), use a ½″ watercolor brush. For larger sheets of paper, choose a brush with a larger head. Buy good-quality brushes for applying sizing, otherwise you will constantly be picking out bits of bristles from the surface of the sized paper.

To apply sizing, first place the paper sheet on a stack of newspapers. Beginning at the top, brush the sizing across the paper from side to side in a long strip. Brush on another strip of sizing, being careful not to overlap the strokes too much. Also be careful not to leave any gaps. When you are half finished

97

with the sheet of paper, hold it up to the light by the dry end to see if there are any spots you have missed. If there are, carefully add more sizing to these areas. Finish the second half of the paper and check it in the same way.

Try not to brush over a part of the paper on which the sizing has already started to dry as this may tear the surface.

Air dry or oven dry the sized paper sheets. For oven drying, the paper can usually be stacked between two couching cloths with no problem of sticking. However, sheets sized with white or animal glue should not be stacked. Dry these with the sized surface uncovered.

After the sheet has dried, the second side can also be sized using the same method.

Sheets sized externally with gelatin and starch can be used with felt-tip markers of all kinds with very little bleed-through. There is only slight bleed-through from watercolor and virtually none from tempera. In addition, pen-and-ink, pastels, charcoal, pencil and ball-point pen can all be used successfully on paper sized this way.

When white glue is added to the sizing brushed onto the paper, the resulting sheet is stiffer, but also more waterproof. Watercolors, tempera and felt-tip markers to not spread or bleed through the paper, neither do acrylics nor pen-and-ink. Of course, pastels, charcoal, pencil and ball-point pen can all be used on paper sized externally with white glue.

Varnish and Shellac. Various varnishes and shellacs can be used to size paper externally. Brush the varnish on one side of the sheet of paper and allow it to dry. This will take several hours or overnight. Then coat the other side. Varnish-sized paper is firm, but still flexible. It is also waterproof and felt-tip markers of all kinds can be used on it without bleed-through. Watercolor does not work well on this paper, though, as the water puddles and dries very slowly. Tempera poster paints can be used successfully, as can acrylics if the paint is mixed with minimal amounts of water. In addition, oil paints, ink, pastels, charcoal and pencil can be used on varnish-sized paper.

THE COMPLETE BOOK OF HANDCRAFTED PAPER

Paper coated with varnish or shellac is not as opaque as paper finished with other sizing. There is quite a bit of show-through, how much depending on the type of varnish used and the thickness of the paper sheet. The result can be almost transparent paper, which may or may not be desirable. Yellow varnish can be used to give the coated paper an aged look.

Spray Varnish. You can also use spray varnish for an easy way to size paper externally.

Shake the can of spray varnish well. Hold it 12″ to 15″ from the paper to be coated (placed in the center of a sheet of newspaper). Spray the paper to coat it lightly, using a side-to-side motion. Let dry about 5 minutes, then apply another light coat of the varnish. Repeat on the other side of the paper sheet for C2S.

Paper sized with spray varnish is still soft and flexible. It is not water resistant and is not suited to use with watercolors. Tempera paints also bleed through somewhat. However, both hard-tip and soft-tip felt markers work well with little or no bleed-through. The hard-tip marker makes a very pleasing fine line on this sized surface. Pen-and-ink can also be used on it successfully, as can ball-point pen, charcoal, pencil and pastel.

Gesso. Gesso is a thin paste originally made from animal-hide glue and a chalky pigment for whiting, applied as a ground to a surface to be painted. Today acrylic gesso is also available. This is usually a thick mixture of acrylic emulsion plus a pigment (titanium dioxide) and an inert chalky ingredient. Acrylic gesso has the advantage over the traditional gesso of remaining more flexible.

Gesso provides a good workable surface for acrylic paints, casein, tempera and even oils mixed with turpentine. Watercolors can also be used but they are not absorbed as readily as they would be by an uncoated paper. Gesso makes the paper waterproof without covering up the natural texture of the sheet, though the natural color of the paper is lost.

To coat a piece of paper with gesso while retaining its surface texture, first dilute the gesso with water. Use a large, soft watercolor brush that leaves no bristle marks. Tack down the sheet of paper you want to coat, and apply the gesso in slightly overlapping strokes. Let the first coat dry thoroughly, then apply another thin coat of the gesso. Two or three coats may be necessary to get the surface you want.

You can also use gesso to *create* surface texture on your paper. Use the gesso thick as it comes from the container and choose a stiff brush. Apply the first coat of gesso in horizontal strokes across the paper. When this first coat is completely dry, apply a second coat of gesso at a right angle. This creates a crisscross canvaslike surface to the paper.

You can make a tinted painting ground by adding acrylic paints to the gesso you apply to paper. Use just a few drops for a faint hint of color. For example, yellow ochre will give the paper a golden tinge. Or you can combine two colors to create a wide variety of hues.

Acrylic Medium and Varnish. Many manufacturers of acrylic paints also offer a matte medium and varnish. This product gives a matte finish when used over acrylic paints. It can also be used as an external sizing for paper.

When applied straight from the container, the matte medium and varnish quickly seeps through the paper to the back side, so place paper to be coated on a stack of newspapers. Using a soft brush, apply the varnish in slightly overlapping strokes. Let dry. If desired, coat the back side of the sheet too.

When applied to both sides of the paper, matte varnish creates a very transparent sheet with a stiff surface texture. It is highly waterproof, however, and tempera, acrylics, water-colors and oils can all be used with this sizing.

Sizing with Color. While you add external sizing with a brush, you can also add color to your paper. Use either acrylics or watercolor paints. Mix a small amount of paint with the sizing and stir well. The result will be lightly tinted paper. For an antique-toned paper, use burnt umber, raw sienna, mars yellow, burnt sienna or raw umber. Be careful not to use too much paint and avoid bright or strong colors as these would hide the surface texture of the paper.

OTHER ADDITIVES

Commercial paper mills add some substances to their paper to increase its dry strength. These include resins and natural gums such as locust bean gum and guar gum. Paper mills also add fillers to increase the brightness, opacity and surface smoothness of the paper.

One such filler is talc and you can add this to your paper too by shaking talcum powder into the pulp. Add two shakes of powder to each blender-load of pulp. If the talcum powder is scented, it will lightly perfume the finished sheet of paper. Talc gives paper a silky texture that is quite pleasing. This is especially noticeable on the smooth side of the sheet.

Whiting. To improve the brightness of paper, commercial mills add chemicals such as titanium dioxide, clay (aluminum silicate), calcium carbonate, zinc oxide, zinc sulfide, hydrated silica, calcium sulfate, hydrated alumina and barium sulfate.

At home you can easily add whiting to your handmade paper by squeezing a bit of white acrylic artists' paint (titanium dioxide) into the blender before the pulp is processed. A 1" long squirt of paint, added to 1 blender-load, is enough to whiten a whole sinkful of pulp. Some foam may form as a result of the paint addition, but this soon evaporates.

Make whiter-than-white paper by adding just a bit of cobalt or ultramarine blue acrylic paint to the blender with the titanium dioxide.

SIZING PROBLEMS

Problem	Cause	Cure
Hairs on paper surface.	Shedding from a poor-quality brush.	Remove hair from sizing if still damp. Use a better-quality brush.

Problem	Cause	Cure
Paper curls.	Coating just one side has created uneven surface tension.	Coat other side of paper with sizing to balance the sheet.
Paper surface tears during external sizing.	Brushing over areas that are already sized.	Be sure not to brush over areas of sizing that may already have started to dry.
Dull areas and spots in paper sheet.	Areas missed during external sizing.	Try to retouch spots with sizing.

7

Special Effects

COLORED PAPER

One of the first changes you may want to make in the basic paper recipe is to add color. There are several different ways to create colored sheets of handmade paper. Try each to see which works best for you.

Colored Pulp

You can start by sorting junk mail into color groups—reds and oranges; yellows, gold and tans; all shades of green; blues, purples and lavender—to make into separate batches of pulp. Tear the junk mail into pieces or use a paper cutter, then cover with water and boil 1 to 2 hours. *Do not add bleach.*

Store the balls of colored pulp separately, too.

This pulp can be used by itself to produce pale pastel sheets of paper or added to a sinkful of white pulp to produce a speckled paper. The colors will be delicate, though, and may not be suitable for all uses. Also, some colored-paper sources, such as old checks, do not break down easily during boiling and some of the dyes used in junk mail may leach out of the pulp during the boiling stage, even when no bleach is added. If this happens, you may end up with off-white paper instead of colored.

Dyeing the Pulp with Fabric Dye

To get really bright colors in handcrafted paper, the pulp must be dyed. The easiest and most economical way to dye pulp is to use a water-soluble fabric dye, available in grocery or drug stores. There are many colors from which to choose and the shade can be varied from light to dark by controlling the amount of dye added to the pulp water. Also, different colors of dye can be mixed together to create new shades.

To dye the pulp for the vat method of papermaking, first prepare a sinkful of pulp and water, but be sure to use warm water since the dye is more soluble at warm temperatures. Then add the dye powder to the pulp water and stir thoroughly. Half a package of fabric dye per sinkful of pulp water will create a deep-colored paper. Use less than half a package for paper that is light enough to use as stationery. To judge color intensity, start by adding a small amount of dye to the water, increasing the amount gradually. The color of the water will be slightly darker than the finished paper.

When the dye is mixed to the desired shade, dip and form a sheet of paper as described in Chapter 4.

Note: Felts and couching cloths used to stack dyed paper may pick up the color of the dye. These should be washed thoroughly before being used again. Even so, the material may still transfer color to future sheets of paper so it is a good idea to keep separate couching cloths for different color batches.

Dyeing pulp for the pouring method. If you are making paper using the deckle box and mold, first add a small amount of warm water to the pitcher you will use to hold the pulp and water after it is blended. Pour the powdered dye into the warm water and stir until dye is dissolved. Blend small amounts (1 tablespoon) of pulp at a time in the blender with 2 to 3 cups of warm water. After the pulp is thoroughly blended, pour it into the large pitcher. Stir after each addition to the pitcher so that the dye is completely mixed with each new batch of pulp water.

After you have blended all the dyed pulp necessary to form a sheet of paper on your mold (½ cup for the mold dimensions given), pour the pulp water into the deckle box and proceed as described in Chapter 4.

Colorburst Paper

A few grains of fabric dye sprinkled onto a wet sheet of paper make a colorful pattern. As the powder dissolves on the wet paper, it streaks slightly, creating interesting designs. Use several colors of dye for a rainbow colorburst.

Colorburst paper can be made by dropping a few grains of fabric dye onto wet sheets of paper. Photo by Chris Miller.

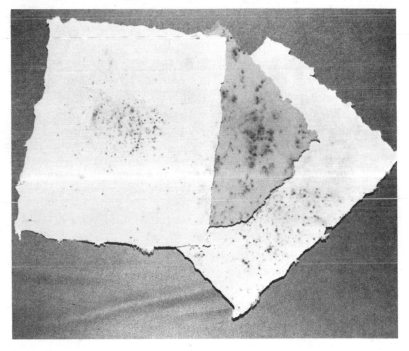

Note: If sheets of paper colored this way are iron dried, the powder granules may make popping noises as they are heated.

Other Ways to Dye Paper

Artists' acrylic colors can be added to the blender before the pulp is processed. Squeeze out a ¼" squirt of paint to give a pastel color to the pulp. Add paint to several blender-loads of pulp for the vat method. Try not to let any paint stick to the sides of the blender while you are adding the color or it will not be mixed in thoroughly.

Blend the pulp and acrylic paint, then pour into the sink for use in the vat method or into a pitcher for the pouring method. Dyeing with acrylic paints is more expensive than using fabric dyes.

Food coloring can also be used to dye pulp, but the colors obtained are not as strong as those from either fabric dye or acrylic paint. Add food coloring to the water in the blender before processing the pulp. The color may fade badly during drying, especially if an iron is used.

Food coloring can also be used to make tint spots or for wet-on-wet painting (see later in this chapter for directions).

Natural Dyes

To create handmade sheets of paper in soft, natural colors, consider using dyes from herbs, plants and flowers. Chances are that there is some plant-dye source around your house or yard right now.

Some plants to look for are: marigold, onion, elderberry, chamomile, blueberry, grape, dandelion, lily of the valley and mistletoe.

You need little in the way of equipment or materials to begin dyeing pulp with plants:

> one or more large kettles: stainless steel, unchipped enamel or glass (brass, iron and copper kettles can affect the dye color)
> pails: plastic, enamel or galvanized iron

stirring rods: use stainless steel or glass rods or wooden
   dowels cut to convenient size (wood absorbs the dye
   color, however, and a different dowel must be used
   with each color)
colander: stainless steel or plastic
measuring spoons and cups
soft water for the dye bath: use rainwater or water that
   has been commercially softened
a record book: some form of notebook to record the
   mixtures and the resulting pulp color

Method. Put the plant parts—flowers, leaves, petals, skins or
whatever—into a large kettle with about 3 gallons of water.
Simmer for 30 minutes to 2 hours. Check the intensity of the
color. When the dye bath is the desired color, remove the kettle
from the heat and let it cool 1 to 2 hours or overnight. Cooling
the dye bath overnight will produce the maximum color
strength. Strain out the plant parts.

Prepare recycled paper as described in Chapter 4. Paper with
a cotton content of 25 percent or higher takes dye much more
readily than wood-pulp paper. Boil the cut-up paper in a
separate kettle for 1 to 2 hours. Drain the cooking water from
the pulp and rinse the pulp until the rinse water becomes clear.

Now add the boiled pulp to the dye bath and return the kettle
to a burner. Simmer for about 1 hour, stirring occasionally.

Drain the dyed pulp in a colander. Then use either the vat or
pouring methods to turn the pulp into paper, or store the dyed
pulp in a sealed plastic bag in the refrigerator.

For fragile plant parts, such as flowers, it is best to put the
plant material in a cheesecloth bag and boil it with the pulp.
This keeps the total boiling time of the plant parts to a
minimum.

Mordants. A mordant is a metallic salt with an affinity for
both fibers and dye. It increases the color obtained from dye
plants. Wool and silk take dyes well but cotton and other
vegetable fibers do not absorb colors as readily. Tannic acid is
sometimes used as a mordant with these fibers.

107

You probably have many mordants around the kitchen right now: alum (potassium aluminum sulfate) is used for canning and is available at grocery stores. It is used to get light, clear colors from the dye plant.

Cream of tartar (tartaric acid) is a common baking ingredient also available from grocery stores. Cream of tartar is often used in combination with alum.

Washing soda (sodium carbonate) is a household cleaner.

Salt (sodium chloride), right from the shaker, can also be a mordant.

Lime (calcium oxide) is used in gardening to sweeten the soil.

Other mordants include stannous chloride (tin), ferrous sulfate (iron), and potassium dichromate (chrome). These are available from chemical suppliers or dye houses. (See Appendix.)

Mordants can be added to the pulp separately or added to the dye bath. Two mordant recipes that work well for cotton and linen will also work for pulp, especially if cotton rag paper is used.

Alum and Soda Mordant

1 cup boiled pulp
3 tablespoons of alum
4 teaspoons of washing soda
1 gallon of water

Alum and Cream of Tartar Mordant

2 cups boiled pulp
1 cup alum
¼ cup cream of tartar
2 gallons of water

To add a mordant to the pulp separately, drain the boiled pulp and rinse it several times until the rinse water runs clear.

108

Put some boiling water in the bottom of a kettle and stir in the mordant until it has dissolved. Add the rest of the water listed in the mordant recipe. Stir in the pulp and bring the bath to a boil, boiling the pulp for about 1 hour. Leave the pulp in the mordant bath overnight. The next morning, remove the pulp and dye it immediately or store it in a sealed plastic bag in the refrigerator.

To add a mordant to the dye bath, stir a small amount of the chemical into the plant liquor after it has boiled. Or the mordant can be added near the end of dyeing if the color is not developing the way you want. Use cream of tartar or stannous chloride to brighten a color; use ferrous sulfate to dull the color.

Experimentation is the key to natural dyeing. You can consult various books on the subject, but many dye recipes are for wool and may not work well with pulp. Try recipes recommended for cotton and linen. Check the reference list in Appendix for books on the subject of dyeing as well as for sources of supplies.

Few plant dyes are colorfast over a long period of time. However, they do fade to attractive, soft colors. To minimize fading, store paper out of direct sunlight.

Be sure to keep a record of the dyes, mordants and types of pulp you use, plus the length of time for the various baths, the date and any other special observations. A notebook is handy for recording mixtures so that a successful dye color can be recreated later. Save a piece of the dyed paper in the record book too.

Plant Materials for Dyeing

Flowers: gather fresh flowers and use them immediately. Simmer the flowers a short time only (10 minutes to 1 hour) because they are so delicate. Or tie the flowers in cheesecloth and boil them with the pulp to cut down the total boiling time.

Leaves: tough leaves should be soaked for 24 hours before simmering. Very tough leaves should even be chopped first before soaking.

Colors

For Yellow: Smartweed (*Polygonum hydropiper*). It grows wild throughout North America in gardens and along roadsides. Use alum as a mordant.

Marigold (*Tagetes* sp.). The flowerheads can be used as a dye source. Use either fresh or dried flowers and add alum as a mordant.

Sumac (*Rhus* sp.). Crush the ripe berries and soak them overnight. Use alum and washing soda as the mordant for yellow. For a gray color, use ferrous sulfate.

Onion (*Allium cepa*). Use the skins from the onions in a 3-pound bag. Add 1½ quarts of water and bring to a boil. Then add 1 teaspoon of alum and paper pulp that has already been boiled and rinsed. Boil the dye bath for about 30 minutes until the desired color is obtained.

For Red: Madder (*Rubia tinctorum*). Soak the roots overnight and use alum as a mordant.

Beets (*Beta vulgaris*). Use either fresh or canned beets. Chop up the root and let it soak overnight. Use alum and cream of tartar as a mordant.

Pokeberry (*Phytolacca americana*). This weed grows over much of the eastern part of the United States, reaching a height of about 12 feet. Use the berries as a dye. Boil these in water with 1 teaspoon of vinegar.

For Blue: Blueberry (*Vaccinium* sp.). Use crushed, ripe berries as the dye source and soak these about 1 hour. Copper sulfate should be added to the dye bath to produce a good color. Try 2 tablespoons of copper sulfate to 2 gallons of water. Use alum and cream of tartar as a mordant.

Elderberry (*Sambucus candensis*). This shrubby tree grows over much of the United States. In the spring the clusters of cream-colored flowers are highly visible along roadsides and these are followed by clusters of dark blue berries. Use the crushed berries plus a teaspoon of salt. Cream of tartar should be used as a mordant.

Larkspur (*Delphinium* sp.). The flowers of this plant can be used to make a blue dye. Chop the flowers and boil them with alum.

For Purple: Dandelion *(Taraxacum officinale).* Put that lawn weed to good use making purple dye. Dig up the plant for the roots. Chop these and soak overnight. Use a mordant of cream of tartar plus alum.

Grape *(Vitis* sp.). Use the dark-colored grapes found in the grocery store. Crush and soak overnight. Use a mordant of alum and cream of tartar.

For Brown: Chamomile *(Anthemis* sp.). Use the crushed flowers from this herb (often available at health food stores) with a mordant of alum and cream of tartar.

Coffee *(Coffea arabica).* Use ground coffee. Boil for 20 minutes, then strain out and discard the grounds. Use a mordant of alum and cream of tartar.

Tea *(Thea sinensis).* The leaves of black tea can be used as a dye source for brown. No mordant is necessary but alum and cream of tartar can be used. Soak the leaves overnight.

For Green: Lily of the Valley *(Convallaria majalis).* The crushed, fresh leaves should be used with a mordant of alum and cream of tartar.

Mistletoe *(Phoradendron serotinum).* This plant is sold around Christmastime. Use the leaves and stems, not the white berries, to make a green dye. Chop up the foliage and boil for about 1 hour. Use alum and cream of tartar as a mordant.

Parsley *(Petroselinum crispum).* Use fresh leaves and chop these. Boil about 1 hour. Use alum and cream of tartar as a mordant.

Spinach *(Spinacia* sp.). Use fresh, canned or frozen spinach. Chop, and boil about 30 minutes. The liquid from canned spinach can also be added to the dye bath. Use alum and cream of tartar as a mordant.

Tint Spots

Tint spots can be used like watermarks to create a unique

111

sheet of paper. The tint spots are actually small areas of the paper sheet dyed a different color.

Acrylic paint, food coloring or fabric dye can all be used to make tint spots, although the food coloring may fade during drying.

There are two ways to add tint spots to a sheet of paper.

Method One. Just before you dip out a sheet of paper from the sink, squeeze a few drops of food coloring or diluted acrylic paint onto the top surface of the pulp water; or sprinkle a few grains of fabric dye onto the water's surface. When you pull the sheet through the colored area of the water, the dye creates a design on the paper. Since the dye or paint spreads on the water, the longer you wait to form the sheet, the larger the colored area on the finished paper will be.

For the pouring method, first add pulp water to the deckle box and stir until there is a uniform mixture. Then add the coloring to the surface of the water inside the deckle box and form a paper sheet.

Method Two. You can also make a tint spot on a formed sheet of wet paper after it has been couched or while it is still on the mold. This technique allows more control of the placement of the design. Be careful, though, for too much dye or coloring can create weak areas in the finished paper which could tear during couching. It is advisable, therefore, to use thick sheets of paper when adding tint spots in this way.

To add a tint spot to a formed sheet of paper, first mix fabric dye or acrylic paint with water, or use food coloring. Add a few drops to the surface of the damp sheet. Use more than one color if desired. The color will spread slightly on the moist paper.

Note: tint spots can stain the couching cloths and felts, as well as the towel used to blot up excess moisture.

Wet-on-Wet Painting. Use fabric dye, food coloring or acrylic paint that has been diluted with water to a flowing consistency. Pour or drop these colors onto a wet sheet of paper, either after it has been couched or while it is still on the mold. Several different colors can be used and abstract designs can be "painted" on the wet paper in this way. Here again, a thick

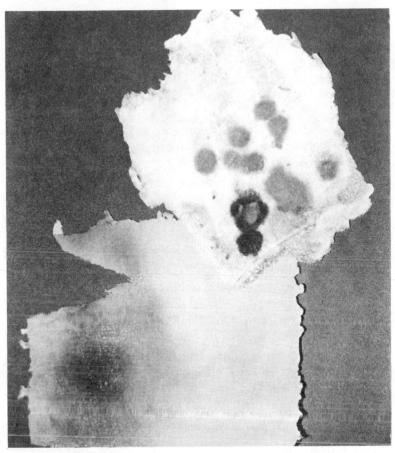

Tint spots made by squeezing a few drops of food coloring on top of the still damp sheet of paper (Method Two). Photo by Chris Miller.

sheet of paper is better able to stand up to the addition of coloring.

Tint Spots from Yarn. Yarn or thread that is not colorfast can also be used to add designs and patterns to paper sheets. Couch a sheet of paper and then place the yarn or thread against it in any design while the paper is still damp. The dye from the thread or yarn will transfer to the paper, bleeding slightly, and will create an attractive colored area.

You can form borders by laying four pieces of yarn or thread along the edges of a sheet of paper. Curl a piece of yarn into a circle or other shape to add a tint spot to one corner of the paper, or form initials with the yarn or thread. You can combine several different colors for special effects.

Rainbow Paper

Beautiful color combinations are possible using different colors of fabric dye or colored pulp. Try both methods of making rainbow paper.

Method One. For this technique, it is really helpful to have another person on hand. Fill the sink with water and blended pulp, in white or any color (see Chapter 4). Then dip the mold to the bottom of the sink and hold it in place. Have a friend stir the pulp water above the mold to get a uniform mixture. Then have the friend add several shades of dye powder, as he or she continues to swirl the water. (Add the dye powder directly from the packages or first pour several different colors into a shallow bowl.) As the dye dissolves in the pulp water, patterns form. Bring the mold to the surface of the sink to form a sheet of paper, capturing these color patterns.

Needless to say, no two sheets of paper made this way will look alike. It is fun to watch the color patterns take shape in the swirling water.

Method Two. Prepare a batch of basic pulp water (white or any color) and fill the sink. Then blend several different batches of colored pulp and pour each into a separate container. When you are ready to make paper, pour the different pulps into the sink and stir lightly. The colored pulp does not disperse in the water as quickly as the dye, so there is more time to dip and form sheets with this method.

Both these rainbow paper techniques can be used with the pouring method of papermaking. Set up the mold and deckle box in the sink and add a basic pulp mixture (white or any color). Then have a friend either pour fabric dye on the water surface inside the deckle box or add the different colored batches of pulp.

114

Make rainbow paper by pouring different shades of fabric dye powder on the surface of the water in the sink (Method One). Photo by Chris Miller.

Another way to make rainbow paper is to add different colored pulp to the sink (Method Two). This sheet was made by this method. Photo by the author.

Marbled Paper

Beautiful, patterned paper with the look of marble can be created with this simple method. You will need completely dry sheets of unsized handmade paper in any color, plus:

> unflavored gelatin
> paint thinner or turpentine
> enamel paint and/or artists' oils in several colors
> a shallow, waterproof container large enough to hold the sheet of paper to be marbled—choose a plastic dishpan, old roasting pan, aluminum-foil pan or photographic tray
> newspapers
> plastic spoons, toothpicks or paintbrushes
> glass jars or old dishes for mixing the paint

Set the container on a counter covered with newspaper. Boil some water and add enough to the container to cover the bottom, then stir in 2 teaspoons of the unflavored gelatin. Mix until gelatin is thoroughly dissolved. Add cold water to within 1″ from the top of the pan. If you are using artists' oil colors, first mix the paint with a few drops of turpentine in a small jar. Stir with a paintbrush or spoon until the paint is runny and flows easily.

Float several drops of the artists' oils and/or colored enamel on the surface of the water in the container. If paint sinks, add more turpentine to the oils or mix the enamel with a few drops of thinner. If the drops of color spread out quickly, dispersing into the water until they are barely visible, then the paint is too diluted. Add more oil color to the turpentine-and-oil-paint mixture; stir the enamel to mix in thicker paint from the bottom of the can. Then add more paint to the water. Adjust consistency until paint floats on the surface.

Using a toothpick, spoon or paintbrush, gently swirl the paint on the water until it forms a pleasing pattern.

Next, carefully lay a sheet of paper on top of the water. The oil-base paints adhere to the paper surface immediately and you

116

Pour the oil-base paint on top of the water-and-gelatin solution, then stir to make a pleasing color pattern. Photo by the author.

Carefully lift the sheet of paper from the water. The oil-base paints will adhere to the paper's surface, forming the marbled pattern. Photo by Chris Miller.

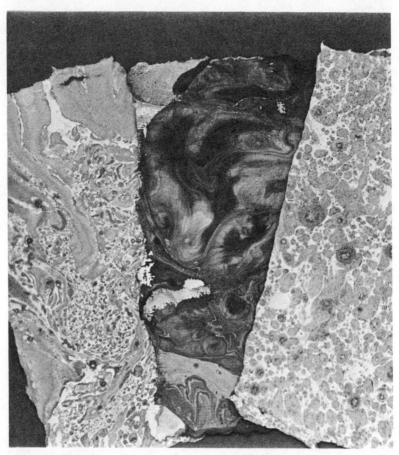
Finished sheets of marbled paper. Photo by the author.

can then carefully lift the paper off the water. Lay the marbled sheet, design-side up, on top of several thicknesses of newspaper. Let drain about 30 minutes, then transfer the sheet to a fresh stack of newspapers. Let dry overnight.

Depending on temperature and humidity, the paper should be dry in about 14 hours. If the paper curls, just place it under a book when completely dry or press lightly with an iron.

You can get a wide variety of different marbled patterns by adding the paint in long streaks across the water surface instead of in drops. Or use a comb to swirl the paint into a design. Drops of paint without any swirling also create an attractive pattern. Try combining two or three colors or several different shades of the same color. Try some natural color combinations for paper that has the look of real marble: pink and gray; pink and pale green plus white; pink, yellow and white; pastel blue and pink; pastel blue, yellow and white and several shades of blue gray plus white. Or use oil-base metallic paints for gold, silver, bronze and copper marbling.

When you are finished marbling for the day or when you want to start a fresh sheet, you can clean off the surface of the water by laying a piece of newspaper on top. The newsprint absorbs the oil paint and can then be discarded.

Marbled paper can also be created by floating oil-base paints on plain water without the addition of unflavored gelatin. The patterns that result from this method tend to be more dispersed but are equally pleasing.

Use marbled paper for notes, greeting cards and even stationery (if the colors are light enough) or use the paper for hand bookbinding. See Chapter 9 for more about these uses.

Metallic Paper

You can give paper a metallic sheen either during sheet formation or after the paper has been formed and dried.

Metallic Paint To create a metallic finish on formed paper, use any of a wide variety of metallic paints available at hobby and craft stores. These come in several colors, including silver, gold, bronze, copper, in bright and antique shades.

To apply the metallic paint, first spread out a section of newspaper to work on. Then shake the paint container well. Using a soft, wide brush, apply the paint to the paper surface with overlapping strokes. From time to time while you are applying the paint, hold the paper up to the light at different angles to check for spots you may have missed. This is especially important if you are coating the toothed or textured side of the paper sheet. Touch-ups added later often show up as darker areas, so it is important to coat the paper completely the first time.

Some metallic paints bleed through badly to the back of the

sheet of paper. To avoid this problem, first coat the paper with gesso, then apply the metallic paint. Using an undercoat of gesso also makes it easier to apply a uniform coating of metallic paint without missing spots. Tempera, acrylic paints and oils can all be used on paper coated with metallic colors.

Gold Foil. Gold foil is an extremely thin, lightweight sheet of metal foil sold in packets at handicrafts stores. It can be added to dry sheets of paper using spray adhesive or varnish.

Spray adhesive is easy to use, but the gold foil must cover the paper completely for any exposed areas of paper that have been sprayed with the cement will remain tacky.

Shake the can of spray adhesive. Apply adhesive to the paper holding the can 6″ to 8″ away from the surface. Spray in overlapping strokes.

Carefully pick up a piece of gold foil. Applying gold foil requires a *very* light touch to avoid ripping or wrinkling the sheet. Lay the foil over the paper where the spray adhesive has been applied. Using a cotton ball, lightly rub the foil to smooth it into place.

Gold foil can also be applied to paper coated with varnish or lacquer. Apply a thin coat of either and allow it to dry until tacky. Then overlay a sheet of gold foil as described.

Gold foil makes a decorative addition to the inside of envelopes. It can also be used to make greeting cards. Since tempera, acrylic paints and oils can all be used over gold foil, artwork or printing can be included in the design.

You also can add gold foil to paper while the sheet is still damp. Lay a sheet of the foil on the paper after it has been couched and lightly press the foil into the damp paper sheet. Then add a felt and press again. Wet paper with foil added can be oven dried.

Bits of gold foil can be added to the pulp water used in either the vat or pouring method. If the gold foil is blended with the pulp, it will be broken down into small, sparkling flecks that blend well with other additives or are interesting by themselves.

Two-Tone Paper

Paper sheets that are one color on one side and a different color on the other are easy to make using double couching.

First make a sheet of paper and couch it. Set this sheet aside but do not add a felt. If the first sheet of paper was white, add dye to the vat now to make the color for the second sheet. Or add dye to change the color from that used for the first sheet: For example, make the first sheet yellow, then add blue to the vat for a green second sheet. If you are using the pouring method, just mix up separate batches of two colors of pulp.

Form the second sheet of paper and couch it too. Hold the couching cloth by the top two corners and line up the second

To double couch, hold the second couching cloth by the top two corners and line the paper up with the sheet that has already been couched. Set the second sheet of paper on top.

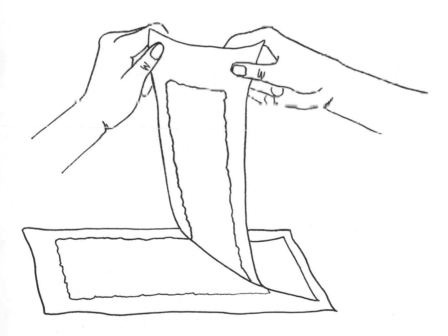

paper sheet with the first one. Try to line up the two sheets as closely as possible, though some overlap is not a problem. In fact, overlapping can be attractive. Lay the second sheet of paper on top of the first. Then place a felt on top of the uppermost couching cloth and pat lightly. Oven dry the laminated sheet at 250° F. for about 20 to 30 minutes. When the paper is dry, it will peel away easily from both couching cloths.

Two-tone paper can be used to make envelopes with a different-color interior. It can also be used for note cards or just as unusual stationery. Besides laminating paper of two different colors, try double couching sheets with different additions: flowers on one side and leaves on the other, for example. (See Chapter 11, Laminated Paper.)

ADDITIONS

Butterfly wings and pieces of yarn, feathers and flowers—all these and more can be added to your handcrafted paper to make each sheet unique and beautiful.

What can you add to paper? The item must be small enough to fit within the dimensions of the sheet. It must also be thinner than the thickness of the paper or close to it in thickness. Bulky additions may cause trouble areas that can tear during couching. The only other limitation on paper additions is your imagination.

Many papermakers find their inspiration in nature—bits of plants, autumn leaves, grasses, seeds, flower petals, even just the stamens from flowers.

Plant material can be easily prepared for adding to paper. To remove the moisture and to flatten the item, use a telephone directory or a large mail-order catalog as a press. Open the book to a center page and slip two pieces of paper toweling between the pages of the opened book, one on each side. Arrange the plant material on one piece of paper toweling as you want it to look when dry. Then carefully close the book.

Set the book aside in some safe spot and place another book or two on top to serve as extra weight. Let the plants dry for at

(Left to right) Paper made with the addition of dried wild flowers, violets and violet leaves and a piece of dried kelp (available at health food stores). Photo by Chris Miller.

least 1 week. Drying may take longer during humid weather. Leaves and flowers are dry when they are no longer limp.

Many bulky flowers can be reduced in thickness by drying in this method. However, very thick flowers such as chrysanthemums or marigolds may be better dried in parts, by pulling off the individual petals and drying them separately. Roses can also be dried in this manner and the individual petals added to sheets of paper.

123

You can also check local florists' shops for dried plants, flowers and attractive grasses. Pull these apart into small sections if they are too bulky to add to the paper whole.

Other natural items that you can add to paper are: moth and butterfly wings, bits of tree bark and feathers.

Butterfly and moth wings are very delicate and require a good hand and some luck to be successfully added to paper. If the butterfly is large, consider adding its wings one pair at a time.

Tree bark must, of course, be in very small, thin pieces and these should not have any curl. If the pieces of bark are not flat, press them under a heavy book for a few days before adding to the paper.

Feathers also make pretty paper additions, but it is generally easier to use small feathers rather than the thicker flight feathers. Look in parks for the feathers of wild birds or ask a friend with a canary, parakeet or parrot to save some feathers for you the next time the bird molts.

There are also many man-made objects that can be incorporated into paper. Consider string, thread, bits of jute, yarn, mohair, embroidery floss, sequins, confetti and glitter. Browse through a local crafts or needlework shop for inspiration, or check around the house for bits of old lace, ribbon, decals or even cancelled postage stamps.

The kitchen provides many paper additions. Check the herb rack for rosemary, ginger, nutmeg, crushed red pepper and bay leaves. Dried herbs can be added whole, like dried flowers, or they can be processed in the blender with the pulp for a few seconds and the crushed herb incorporated into the paper sheet.

All these items and more can be added to paper in any one of three ways: pulp, sheet or vat addition.

Pulp Addition

In this method, the material to be added is carefully stirred into the pulp and water after they have been blended. Then the whole mixture is poured into the sink or directly into the mold and deckle box.

This sheet of paper was decorated by sprinkling glitter on top of the damp sheet. Photo by Chris Miller.

Pulp addition allows the added material to become well coated with paper fibers so that it becomes an integral part of the paper sheet. But the method is haphazard: there is little control of the exact placement of the addition. Pulp addition works well where a random or overall pattern is desired.

Good choices for pulp addition are: small dried flower petals, a handful of crushed herbs, sequins, confetti, many pieces of yarn, several feathers or bits from the tops of grasses.

125

Sheet Addition

This method allows the craftsperson more control over the precise placement of the addition. First a paper sheet is formed either by dipping the mold into the sink or by pouring pulp into a mold and deckle box. When the paper is formed and before it is couched, the additions are carefully placed on the wet surface of the sheet.

This method works well with very thin additions. Bulkier items may not adhere securely to the paper surface and could come loose later.

The sheet addition method is well suited to small items, such as glitter, confetti or spices. Just sprinkle these over the wet paper. Let sheet set a few minutes before couching.

Be sure your hands are dry as you work, for drops of water from wet hands could cause blemishes in the damp sheet of paper.

Vat Addition

Items can also be added to the water in the vat directly above a sheet of paper as it is being formed on the mold. When added this way, feathers, postage stamps, thread and yarn form interesting patterns that are hard to reproduce by hand (that is, if added to a finished sheet). Also, items added this way get a thin coating of pulp and become a more intrinsic part of the paper sheet than they would if placed on top of the formed paper.

Vat addition can also be used with the deckle box in the pouring method. Simply pour in the pulp and, as you raise the deckle box and mold unit to the surface of the tub or sink, drop in the additions. The suction formed will draw the items into the paper sheet.

Scented Paper

Perfume added to the pulp water often loses its fragrance when heat is applied to the sheet of paper with the iron or in the oven during the drying process.

So, to make scented stationery, add perfume or cologne to

126

paper that has already been dried. Check a test piece of paper first to be sure the fragrance will not stain the paper. Store the paper in a tightly-covered box to preserve the scent.

EMBOSSED PAPER

You can easily add texture to handmade paper during iron drying. Just make a sheet of paper by either the vat or pouring method, then couch it with a regular cloth and stack with a felt. Pat the felt lightly to remove as much excess water as possible. Then take the couching cloth with the sheet of paper still attached and place it on a wooden cutting board.

Set up for iron drying (see Chapter 5) but instead of using two smooth-textured couching cloths to dry the paper sheet, choose for one of them a piece of fabric with a prominent texture. Open-weave dish cloths can be used, as can double-knit fabrics with a raised-pattern design, or use burlap or canvas. Check your local fabric stores for suitable materials.

The textured fabric serves as a mold for the wet paper. While you press the paper between the textured material and the other couching cloth, a reverse image of the pattern is embossed onto the paper.

When the sheet of paper is dry, remove it from the couching cloth. It will be embossed on at least one side, possibly both sides—if the paper is thin and the raised pattern of the fabric quite pronounced. Set the finished sheet of paper under a light book or a magazine to keep it from curling. (Do not use a heavy book as too much weight could flatten the pattern.)

For a more pronounced pattern, use two pieces of textured fabric.

Embossed paper can be used for artwork, notes, cards or envelopes.

Embossing with Objects

You can use the principle of embossing to add spot patterns to sheets of paper. The shape of any small, flat object set between a sheet of damp paper and a couching cloth will become embossed into the paper as you iron dry the sheet.

Three sheets of paper embossed with three different pieces of fabric (fabrics shown at top, paper at bottom). Photo by the author.

Sheet embossed with a key that had first been coated with chalk dust.

Look for items like keys, coins, medals, twigs and leaves or use anything that is relatively flat and that will not melt during ironing. Leaves are especially good choices for embossing because the veins form such intricate patterns. Embossed patterns can be subtle and may not be noticed in a sheet of paper until it is turned toward the light.

If you want a more dramatic embossing effect, you can add color. To make the embossed and colored key shown here, the key was first coated with a felt-tip marker. Then pastel chalk dust was sprinkled on top. When the damp paper was pressed with the key in place, the paper absorbed the color with little spreading and the shape of the key was also embossed into the sheet.

Watermarks

Watermarks were first used by Italian papermakers in the late thirteenth century and no one is quite sure why. Some theories are that the designs were used to identify the papermaker—as a sort of trademark—or that the designs identified size, or simply that papermakers found watermarks decorative.

Whatever the reason for their use, early watermarks were simple designs such as crosses, circles, triangles and knots. These forms were twisted in coarse wire and then sewn onto the screen of the paper mold. Often the attaching stitches were also visible in the watermark produced in the sheet of paper. As time went on, the designs used for watermarks became more elaborate and diverse, with finer wire used.

The hand was a popular early watermark, symbolizing fidelity and labor. The unicorn was another common design and stood for purity and innocence. Researchers have recorded over 1,000 examples of the unicorn watermark in the sixteenth, seventeenth and eighteenth centuries. Other popular designs included mermaids, angels, saints, animals such as dogs and bulls, and coats of arms.

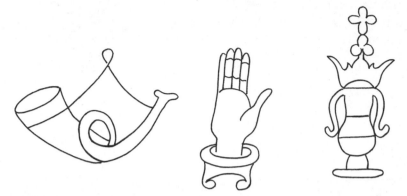

(Left to right) The hunting horn, hand and a decorative flagon were all used as watermark designs.

The dog (right) and the unicorn were also popular watermark designs.

The bull was a common watermark during the 1300s, as were flowers and the rooster.

In early America, watermarks generally consisted of names and initials. The first paper mill established in the colonies by William Rittenhouse in 1690 had as its original watermark just "Co." to stand for company. Later, a dove holding a branch in its mouth became a favorite American watermark.

How does a watermark work? The raised design of the mark on the mold screen causes the layer of pulp formed on top of it, and hence that area of the finished paper, to be thinner. When the paper is held up to the light, more of it shines through the thin watermark area, outlining the design. The design becomes a permanent part of the paper sheet and cannot be removed without destroying the paper itself.

When watermarks are made by a Fourdrinier machine, the dandy roll embosses an impression on the wet sheet of paper, after it has already been formed. Because of this, machine-made watermarks are not usually as sharp as those made on hand-dipped paper where the watermark is actually formed during the papermaking process.

HOW TO MAKE A WIRE WATERMARK

Watermarks used to be made out of thin brass or silver-plated wire sewn onto the wire mold-screen with even finer wire. Today, jewelry-making wire can be used instead, with cotton or polyester thread to sew the design to the fiberglass-mesh mold screen.

Purchase fine wire at a jewelry or crafts store. The wire should bend well without forming a sharp angle or breaking. It is easier to sew a watermark to a screen before it is attached to the mold.

To make a watermark, sketch a design on a piece of scrap paper. Then, using needle-nose pliers, twist the wire to fit the design. Do not cross the wire over itself because pulp will collect in these spots. Instead, cut the wire where it would cross and sew the pieces to the screen separately.

When you have shaped the wire to match your drawing, set it on the screen where it will appear on the final sheet of paper.

133

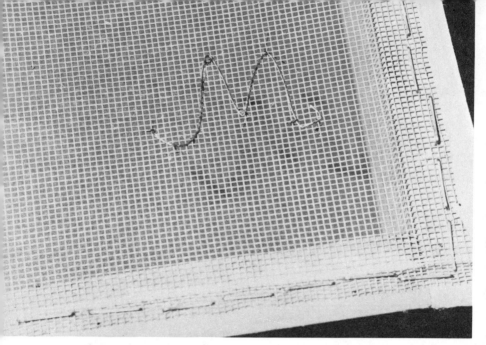

Finished watermark sewn to fiberglass-mesh screen.
Photo by Chris Miller.

A sheet of paper formed on the mold with the watermark shown in previous photograph. Note also the pattern from the mesh screening. Photo by the author.

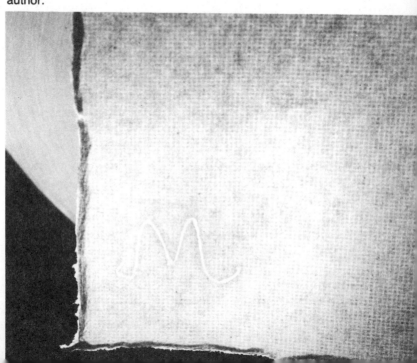

Remember to allow for the width of the mold edges. Cut off a piece of polyester or cotton thread and begin tacking the wire design to the fiberglass screen. Loop the thread over the wire wherever it forms a curve, bring both ends of the thread to the back of the screen and tie in a double knot. Continue tacking the design onto the screen until it is held firmly in place.

Though watermarks have traditionally been small and set in a corner of the paper sheet, you can also use larger, central watermarks as the main design element in artwork. Use heavier wire for the design to be more visible.

LACE WATERMARKS

You can use pieces of lace to make a watermark design on handmade paper sheets. Tack the lace to the mold screen using polyester or cotton thread as described. Run the lace off the edges of the mold. If an edge is left in mid-screen, it must be tacked down carefully or the lace edge can make a hole in the finished sheet of paper.

THREAD WATERMARKS

These designs are very easy to make and are recommended for use when teaching papermaking to large groups and to children.

All you need is embroidery floss and a large needle. Use colorfast thread or the design will bleed onto the damp paper.

Thread the needle using unseparated embroidery floss. Knot the other end of the floss with a knot large enough so that it does not pass through the openings in the screen mesh.

To make initials composed of straight lines, A, E, F, H (and so on), use one long stitch for each line in the initial. For example, to make the letter M use four separate stitches. Begin the letter at the base of one of its legs. Pull the thread through the mesh toward the front of the screen until the knot catches. Make a straight stitch and pull the thread to the back of the screen again (see illustration). Come up again just to one side of the end of the stitch, but through a different hole in the

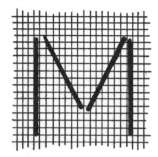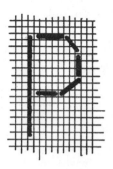

Diagrams for making initial watermarks with embroidery thread.

screening, and make a diagonal stitch for half the middle section of the **M**. Make another diagonal stitch, as shown, then complete the letter by adding another straight stitch. Pull the embroidery thread tight to the back of the screen and knot it off.

To make initials with curved lines, use straight stitches for all the lines you can. For example, to make the letter **P**, use a straight stitch for the back of the **P**. At the top, make another straight stitch out to where the top of the **P** begins to curve. Then pull the thread to the back of the screen and make a series of small, straight stitches to form a curve. (See illustration.) Add another straight stitch to the bottom of the curve to complete the letter.

You can also use embroidery thread to make watermark designs other than initials—try crosses, shields, triangles and so on.

The Embroidery Hoop Mold makes it possible to have several screens, each with a different watermark attached. Change your watermark to suit your mood.

A sheet of round paper formed on the Embroidery Hoop Mold. This sheet has a watermark formed by the thread method (in the lower right-hand corner) and also has a cutout design made with an embroidered star appliqué. Photo by the author.

CUTOUT DESIGNS

Fabric and notion stores carry a wide range of embroidered appliqués for use as decoration. There are stars, anchors, hearts, flowers and initials. When these appliqués are tacked down to a mold screen, they act as a die that forms a cutout area in the finished sheet of paper.

To make a cutout design, choose an appliqué that has a clear, easy-to-identify shape and one that is thick. Choose simple shapes, such as initials, stars and so on. Elaborate designs may not show up well. Tack the appliqué to the screen mesh using cotton or polyester thread.

Form the sheet of paper. There should be little pulp on top of the appliqué when you finish. Couch the sheet of paper (see Chapter 5). If the appliqué is thick enough, there will be a clear cutout design left in the paper sheet.

Use a cutout in a corner of a sheet of writing paper or incorporate one or more into the design of greeting cards or notes. Appliqués can also be used to make colored watermarks.

COLORED WATERMARKS

The process for making colored watermarks was developed by Sir William Congreve during the early 1800s. His method was to make a thin sheet of white paper and couch it. Then a colored sheet of paper was formed in which a design was stenciled. This colored sheet was couched on top of the white sheet. Then a third sheet, white again, was dipped and couched on top of the colored paper, forming a triple layer. When held to the light, this laminated paper showed the design cut in the middle layer.

You can make colored watermarks in the same way today and it is not difficult to do the double and triple couching involved.

Stencils for the design in the middle sheet can be made by tearing the paper after it has been couched. However, shapes made this way are somewhat crude. You can also use an embroidered appliqué as a die to cut the stencil. Forming a sheet of paper over one of these appliqués gives a clean-edged cutout area that will provide the design in the middle sheet.

To make colored watermarks, first form the two outer sheets of white paper. Couch and set aside. Do not stack them or add felts. Add dye to the vat to create the colored middle sheet (or use a pitcherful of dyed pulp with the pouring method of papermaking). Be sure to use nonrunning dyes so that the color does not spread to the outer layers of paper in the laminated sheet.

Using a mold with one or more embroidered appliqués tacked to the screen, form the colored sheet of paper. Couch it. Holding the couching cloth by the top two corners, line up the colored paper sheet with one of the white sheets you have already couched. Starting at the bottom edge of the couching cloth, carefully place the cloth and colored paper on top of the white sheet. Pat in place with your hand, then carefully remove the couching cloth. Pick up the couching cloth containing the other outer sheet and lay it on top of the double-couched paper in the same manner as described above. Pat gently and this time leave the cloth in place. Oven dry at 250° F. for about 20 to 30 minutes. When the laminated paper sheet is completely dry, remove it from the couching cloths and hold to the light. You will be able to see the design in the middle sheet.

More than three layers can be used to make colored watermarks, with several different colored sheets sandwiched between the outer layers. Make a different stencil design in each sheet, placing them at different locations on the mold screen.

9

Paper for Special Purposes

ART PAPER

One of the main qualities sought in art paper is permanence. Permanence depends on the raw materials used in the pulp, the most permanent papers being those made from pulp which is as close to pure cellulose as possible. These papers resist yellowing and are stronger.

Cotton and hemp are almost 100 percent pure cellulose, while wood pulp is only 50 percent to 90 percent cellulose. The cellulose content of wood pulp is not high enough for real permanence and the paper made from this pulp begins deteriorating in a much shorter length of time than does rag paper. Paper permanence also depends on pH. Acid deteriorates paper, so it is important to keep the pulp at a neutral pH.

Of course, practice art paper can be made from wood pulp (recycled junk mail). If you are interested in making your own art paper, spend some time browsing through an art store. Buy small quantities of different types of paper and try them out to see which finishes, colors, sizes and thicknesses you prefer. When you have an idea of the type of paper you enjoy using, you can begin making your own.

Watercolor Paper. Watercolor painters prefer a heavy, 100 percent rag paper. To get a heavy sheet that will stand up well to repeated wetting, use a larger proportion of pulp to water when making the paper. (But do not blend more than 1 tablespoon of pulp at a time.)

Both rough- and smooth-surfaced watercolor papers are used, the finish depending on the preference of the artist. However, a rough surface is usually preferred. This texture can be achieved by drying the damp paper couched on a piece of burlap. See the directions for embossing in Chapter 7.

Handmade and machine-made watercolor papers are sold in art stores, but handmade is chosen by most watercolor painters because they like the "character" of the paper. When trying to explain this term, watercolor painters usually mention that the slight irregularities of handmade paper make it more challenging to use.

Sizing is also a matter of preference. Water leaf paper absorbs the color rapidly. This type of paper is used by artists who work rapidly and make few changes—for the paper allows little time to rearrange the colors. On the other hand, some artists prefer hard edges in their paintings or they want a chance to change their work. A less absorbent, sized paper allows both. Use an internal or external gelatin-and-starch sizing to prepare this type of watercolor paper. (See Chapter 6.)

Pastel and Charcoal Paper. Artists who work in pastels also prefer 100 percent rag paper because of its durability. The tooth and color of the paper are matters of individual taste.

Tooth is the texture of the sheet surface. Pastel paper can have a fine surface texture (fine tooth) or a rough one. Using a mold with a fiberglass screen will produce a paper with a good degree of tooth on the side formed next to the mold. Since brass screen is much finer, paper made on it will have a smoother surface.

Colored paper can be used with great success for pastel drawings. Usually, bright, strong colors, are not recommended for use with pastels; instead ivory, tan, gray or other neutral shades are chosen. Use commercial or plant dyes to tint pastel paper. Either water leaf or sized paper can be used.

Charcoal paper is generally fine toothed and has a pattern known as "laid lines" (from the laid mold on which it used to be made). Today the pattern is given to machine-made paper by the dandy roll of the Fourdrinier machine. You can duplicate this pattern in handmade charcoal paper by using a laid mold. (See list of suppliers in Appendix.) Charcoal paper is usually white and either water leaf or sized paper can be used with this medium.

Drawing Paper. White is the most popular color for drawing paper though an antique or off-white can also be used with good results. The drawing paper should have a slightly toothed surface, the degree of texture depending on the preference of the artist. Use either water leaf or sized sheets for pencil drawing.

Paper for Acrylics. Acrylic paints offer the artist the choice of using them like watercolors (mixed with a good amount of water and applied to the paper in washes) or like oils (with little or no water added and the paint applied thickly, perhaps even with a palette knife). The type of paper used will be determined by the method of painting.

If acrylics are used with a large amount of water, choose a watercolor paper—one with either a smooth or rough surface, water leaf or sized.

However, if you plan to use acrylics undiluted, you will need a different type of paper. Make a thick sheet and use a sizing—starch, gelatin, a combination, or even varnish or gesso. You can get a canvaslike texture by embossing wet paper with a sheet of canvas. (See Chapter 7.)

Paper for Oils. While most oil painting is done on canvas, sheets of properly-sized paper can also be used for practice.

Make a thick sheet of paper and emboss it with a canvas pattern. The paper must be sized externally with varnish on both sides to take the oil paints and stand up to the turpentine or thinner, or it must be coated with gesso or acrylic matte medium and varnish. After the sheet of paper is sized, you may want to back it for extra strength by attaching it to another sheet of paper with glue.

Decoupage. Handmade paper, with its antique charm, lends itself to use in decoupage. Choose delicately frayed sheets of paper as backgrounds for decoupage prints, pictures and decals. Use sized sheets for decoupage work since water leaf absorbs more of the coating.

Paper for Making Rubbings. Making tombstone rubbings has become a popular craft but there are many other interesting objects that can be used to make rubbings: old coins, medals, plaques, manhole covers, historic monuments, ornaments on old buildings and even plants—leaves, flowers or fern fronds.

A rubbing can be made of an entire tombstone, plaque or other large building ornament, or you can make a rubbing of just one part.

The clearest rubbed image comes from a recessed design, one where the design has been carved into a flat surface—which explains why tombstones are classic subjects for the craft. However, some raised designs also produce clear rubbings— often even the raised lettering on a coin will come through. To check a design for suitability, use a piece of newsprint to make a test rubbing.

To make a rubbing, you will need: a small cake of finishing wax (available from shoe repair shops or wholesale leather companies—or see the list of suppliers in Appendix) or a glass-marking pencil (available at art supply stores) and, of course, paper.

Make a thin, lightweight paper for doing rubbings. It should be about as heavy as bond typing paper. Water leaf is fine. Any color can be used, but an off-white or antique cream color are most effective, especially for historical rubbings. Be sure to use the smooth side of the paper to make the rubbing since the tooth of the textured side can interfere with reproduction of small details in the image.

Other supplies that are helpful, especially if you are making a field trip to look for designs: masking tape (to hold the paper in place); a folder, art folio or mailing tube (to protect the finished rubbing on the way home); a brush (to clean away dust or moss) and newsprint (for test rubbings).

How to Make a Rubbing. The key to getting a good impression is to take your time. Center the paper over the design to be used. If possible, use masking tape to hold the paper in place. To make a rubbing of a small object, tape the object to a counter or table, then tape the paper sheet on top.

Make smooth, even strokes with the wax or pencil—there should be no lighter or darker areas. Also, be sure to make the strokes all in the same direction. It's a good idea to start with light strokes; these can always be darkened later. (A medium gray background is usually preferable to black, so do not darken the strokes too much.)

Rubbings made from small objects or plants can be used to make stationery or greeting cards. Large rubbings can be matted and framed as interesting works of art.

Mats. Mats used to frame paintings, photographs and prints may be made from mat board that ranges from thin and flexible to fairly thick and rigid. You can make attractive mats from handmade paper.

The best paper to use for mats is 100 percent rag paper. It is the most permanent and will not stain or discolor the artwork. If permanence is not an important factor, then any pulp can be used to make the paper for the mats.

Use a mold that is sufficiently larger than the item you want to frame and make a fairly thick sheet of paper. A mat of thick handmade paper will not buckle easily. Make the paper any color you think would enhance the artwork or photograph to be framed—black, white, off-white, gray, beige or any color at all. You can also add texture to the mat by iron drying the paper sandwiched with a piece of textured fabric or burlap. (See Chapter 7.)

To cut the mat and frame the artwork, you will need:

> a razor blade, or a sharp mat knife (available from art supply stores)
> a metal straightedge or ruler
> gummed tape (for attaching the artwork to the mat—use a good-quality gummed paper, not masking tape, clear tape or gummed brown wrapping tape since all of these can discolor the artwork)

Measure the frame you plan to use to determine the outside dimensions of the mat. Mark these measurements on the back of the mat paper. Working on top of several thicknesses of newspaper or a piece of cardboard, use the straightedge as a guide to cut away the excess mat paper with the razor blade or knife. Use a light, even pressure on the blade.

Next measure the artwork to be matted to determine the inside dimensions of the mat. Mark off these measurements on the back of the mat paper using a pencil and ruler. Place the artwork over the proposed opening to check your measurements. Make any corrections. You are now ready to cut the mat.

Using the straightedge as a guide, cut along the pencil lines. For a beveled edge, hold the knife or razor blade at an angle. If necessary, lightly sand the cut edges of the mat.

Center the mat over the picture to check the opening size. Make any necessary adjustments. Carefully turn the mat and artwork over as a unit and attach a strip of gummed tape across the upper edge of the artwork. Do not fasten the artwork to the mat along all edges because this would allow no room for the art paper to move as its moisture content changes from day to day. Buckling could result. The matted photograph or artwork is now ready to frame.

100 Percent Cotton Rag Paper

If you want your artwork to last with minimal deterioration, it should be executed on 100 percent rag paper and should also be framed with rag mats and mounted with high-quality gummed tape.

You can make 100 percent cotton rag paper at home even if you do not own a beater to macerate old rags. There are several methods:

1. You can recycle 100 percent rag paper. Use either high-quality stationery, rag mat board or watercolor paper. Save bits and pieces of the paper or recycle ruined paintings. Tear the paper into small pieces, boil and blend as recommended for recycling junk mail. (See Chapter 4.)

2. You can purchase linter fiber from a hand paper mill. This fiber can be prepared in a home blender. Linter fiber is the seed hair of the cotton plant. It is usually available in a dry sheet form, with each sheet weighing about 10.57 ounces. Paper mills frequently require a minimum sheet order, but the price per sheet is very reasonable.

3. Many hand paper mills also sell 100 percent cotton pulp. This pulp is ready to use; it does not even need to be blended.

The pulp is usually sold in a liquid form, just as it comes from the paper mill's Hollander beater. Pulp is most often 100 percent cotton and of a neutral pH. The price for pulp is very reasonable, but shipping charges for the usual 5-gallon containers can be expensive. However, if you live close enough to a hand paper mill, you may even be able to return the containers for refilling.

Usually, mills sell only white, unsized pulp. Some, however, offer custom coloring and sizing for an additional charge. Often, too, the paper mill will beat the pulp to your specifications, working with you closely on your project. Write to mills listed in Appendix for more information.

BOOKBINDING

The first books consisted of long papyrus scrolls, wound onto wooden poles. The modern book began to take form around the second century A.D. in Greece when sheets of parchment were bound together. In the Middle Ages, bookbinding developed into a fine art. Monks fastened sheets of vellum together along one side and added protective coverings that were sometimes ornamented with gold, silver and precious gems. With the invention of movable type, handmade then machine-made paper became the most common page material used in books.

Whether the hand papermaker becomes interested in bookbinding or the bookbinder is drawn to papermaking, the one craft truly complements the other. There are few professional hand bookbinders in the country today, but the craft is enjoying increased popularity among amateurs.

There are basically two aspects of hand bookbinding: restoration of old books and the creating of original, limited-edition books.

Restoring old books is the area of bookbinding with which most professional craftspeople are involved. Any torn or damaged parts of the book to be restored must be replaced with materials as close as possible to those that would have been used at the time the book was made. This includes handmade paper. Even paper flaws are retained during restoration, e.g., when a book is pressed, the bookbinder is careful not to flatten out the ripples in the paper that resulted because the sheets were not dry before they were printed.

Handmade paper is needed not only for the pages of restored books; it is also a necessary part of the binding itself. Endpapers are decorated sheets at the front and back of the book, used to cover the edges of the leather binding. Endpapers developed into an art form in themselves, and beautiful hand-marbled papers are available still from Europe. The hand papermaker can marble paper for this purpose using the directions given in Chapter 7.

The other aspect of hand bookbinding—creating original, limited editions—is a highly creative art. Journals, art sketch books, handprinted volumes of poetry, folios of watercolors, diaries, wedding albums, visitors' books and address books can all be handbound. Handmade paper can be easily adapted to each project in regard to color, finish, fiber type and size of sheet.

For more information on hand bookbinding, see Appendix which lists books on the subject as well as suppliers of materials and instruction.

STATIONERY

A very practical way to use handmade paper is to turn the sheets into stationery—letter-size sheets, notes and even envelopes. Using some of the easy printing methods described later in this chapter, you can also create original holiday and greeting cards.

Writing Paper. Browsing through a good stationery store will give you many ideas for creating your own unique stationery. Add bits of dried flowers to one corner of the sheet, add a small arrangement of dried grass, your own watermark or make beautifully colored paper.

You can make borders for your writing paper using yarn or thread that is not colorfast (see Chapter 7), or try stick painting, described later in this chapter.

An interesting edge can be given to paper sheets by using a mold without a deckle. Dip the mold in the sink and form a sheet of paper. The pulp will spread out to the edges of the mold, even over the wooden sides. Using a blunt table knife,

Make a raised edge to sheets of paper by using a blunt knife to scrape the pulp in from the sides of the mold about ¼" or so. This also makes couching the sheet easier. Photo by Chris Miller.

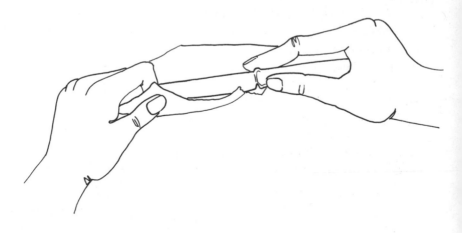

To tear paper, retaining a ragged edge, hold a blunt knife against the paper sheet.

gently scrape the pulp back from the edges about ¼" or so, in toward the center of the mold. Then couch the sheet in the usual manner. The edge that results will be nicely irregular and thicker. (Many of the sheets of paper shown in this book were done this way.)

Store-bought stationery comes in standard sizes: 8½" x 11" for business; 5⅞" x 7⅞" and 6⅝" x 7⅞" also are common. When you make your own paper, there are no restrictions on the dimensions of the stationery you create, except for post office regulations, since you will also be making the envelopes to fit.

Let sheet size be determined by the size of your mold or adjust the sheet size: tear paper in half or thirds to get the dimensions you want. Make a larger or smaller mold for special stationery.

To keep a naturally ragged edge while tearing paper, hold a blunt table knife against the paper and tear (see illustration). The knife keeps the tears from ripping too far into the sheet.

150

If you want a quantity of stationery that appears the same in color, texture, pulp quality and so on, be sure to form it all at one time. Recycled paper pulp varies from batch to batch so that it might be difficult to exactly duplicate a sheet at a later date. If you are using dyes to color the sheets, this is even more important. Paper from different dye lots will differ in color.

So, make up a number of sheets of stationery at one time— fifteen or twenty sheets—and if you want the envelopes to match the paper, then make the paper sheets for these at the same time too.

Envelopes. To make envelopes, first measure the width of the sheet of paper you plan to use as stationery, then allow an extra ½" for the envelope width. Use the envelope pattern included here or carefully pull apart the glued flaps of an envelope you have and use it as a guide.

Pattern for an envelope.

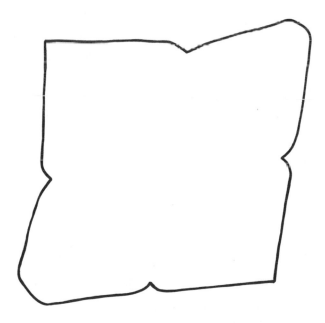

Enlarge the envelope pattern to the desired size, cut it out and set it (or the envelope you have opened) on top of a sheet of handmade paper. Lightly trace the outline of the envelope with a pencil and cut out.

Note: If you want any of the envelope flaps to have ragged edges, be sure to cut the paper slightly larger around this section. Then, using a blunt knife held against the paper, carefully tear the edges.

Fold in the three bottom flaps of the envelope (see illustration) and glue in place with white glue. Let dry. To use the envelope, either glue the flap down with white glue after you have enclosed a letter, or use sealing wax. Sealing wax is sold in stationery stores and comes in a candlelike stick. Light the wick

Make the envelope by folding the flaps in the order numbered.

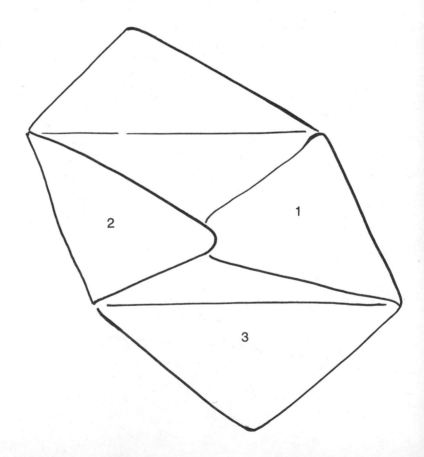

Make note cards by folding them in one of two ways. View One on left. View Two on right.

and hold the wax over the closed flap of the envelope. Let several drops of wax fall onto the flap until a small puddle is created. Use a coin, ring or seal (also sold in stationery stores) to make an imprint in the wax.

Notes and Greeting Cards. Notes can be folded two ways. (See illustration.) To make notepaper, first fold the paper sheet in half. You can then print the front of the card, using one of the printing techniques described later in this chapter. Or perhaps you've added pressed flowers, spices, yarn, feathers or other decorations as the paper was formed. (See Chapter 7.)

Note: If you plan to use the card as shown in View One, be sure that any item with a definite top and bottom (such as a flower) is added to the right half of the paper sheet as it is

153

formed on the mold, or that it is printed on the correct page of the note. (See illustration.) To make a note card as shown in View Two, the design can be added to either half of the mold, to either outside page.

You can create all types of cards—birthday, get well, holiday, congratulations and Christmas—using either of these note-paper folds. Use one of the printing methods described in this chapter to decorate the cards or draw your own designs with felt-tip markers. You can even make your own humorous cards to fit any occasion—keep a supply of blank cards on hand and when a special situation comes up, you can create a card to fit. See Appendix for books on designing your own greeting cards.

You can also make personalized cards featuring your favorite quotes, sayings or poems. Hand letter the verses onto the card using special lettering pens, available. at art supply stores. There are many books on the subject of hand lettering and calligraphy. (See list in Appendix.)

Another idea for handmade note or greeting cards is to use a theme, such as a collection of recipes with illustrations, signs of the Zodiac, flowers of the month (added to the pulp). Many craftspeople tie the theme of the cards to their own area by featuring local recipes, sketches of landmarks or historical buildings or by incorporating pressed local flowers. These theme cards make nice souvenirs for tourists and usually sell readily at local crafts shops and bazaars.

Stationery, envelopes and cards also make nice gifts. Make a watermark of the recipient's initial or use his or her favorite color or flower.

Invitations and Announcements. Using one of the card folds, you can make small notes to serve as invitations for parties, open houses and showers or for birth announcements; smaller cards still, can serve as bridge tallies and gift enclosures. Use a lettering pen to hand letter the information.

Even wedding announcements can be made at home and these are especially attractive when created from an antique cream-colored paper with delicately frayed edges. If you are not as good at lettering as you would like to be, check for art students in your area who could do the job for you.

Packaging Stationery

To package stationery or notes for sale or gifts, you can use a simple method that does not involve finding an appropriately-sized box.

Cut out a section of thick plastic wrap about 3" larger all around than the cards or writing paper you are wrapping. Spread the plastic out on a flat surface. Stack the stationery in a neat pile then place it face down on the plastic wrap. Fold the plastic to the back, pressing the flaps securely in place. Add a piece of clear tape across the last flap to hold the package together. Next use a pretty ribbon, piece of lace or yarn and wrap the package, tying a bow in front. Add a flower, piece of dried grass, candy cane, etc., for a festive look. If the paper is to be sold, you can write the price on a self-adhesive unprinted label (sold at office supply stores) and stick this to one corner of the package.

BOOKPLATES

Bookplates are a charming, old-fashioned custom it is fun to revive, especially when you can create your own from hand-made paper. Bookplates are also a good item to sell at bazaars or local crafts shops.

Use any color handmade water leaf paper cut or torn to a 3" x 4" size. Add a printed design or use a piece of paper with your own watermark in the center. Write or print your name on the bookplate or leave a space for the name of the book's owner.

With a flat brush, apply a very thin coating of library paste to the back of the bookplate, then attach it to the inside front cover of the book. Lay a couple of pieces of typing paper over the bookplate to act as blotters, and support the open cover of the book by placing one or more books or magazines under-neath. Place another heavy book on top of the typing paper as a weight while the paste dries. This should take a few hours, depending on the weather.

CALENDARS

Make your own personal calendar with hand-lettered dates

and special occasions or give a handmade calendar as a gift to a relative or friend.

To make a calendar, use any size paper in any color on which the dates will be easy to read. Print a design on the top half of the paper or add ferns, flowers or feathers to the top half of each sheet as it is made.

Next, letter the names of the months on the calendar pages and add the days of the week. You can use a hard-tip felt marker with good results. Make five rows of seven boxes each on every page. Draw these lines on the calendar or if the paper is thin enough, use a grid placed underneath the page as a guide for adding the dates. Next mark special birthdays, anniversaries and holidays on the calendar. You can even add special sayings or proverbs.

Using a hole punch, make a hole in the same position on each side of the top edge of each page of the calendar. Set the pages in order by month and thread a piece of yarn or ribbon through the holes, then tie in a bow. Your calendar is now ready to hang.

If you plan to make many calendars to give as gifts or if you want to sell specialized calendars at local crafts shops or church bazaars, consider making small linoleum block prints for the numbers 0 to 9. Using these to print the dates not only saves a great deal of time, it also makes each calendar page more uniform.

PLACEMATS

You can easily make unique waterproof placemats using your favorite sheets of handmade paper. Use paper that is colored, marbled or that has special additions. Choose one of three methods for making placemats.

Method One. For this method you will need clear, self-sealing plastic sheets. These are available at stationery and office supply stores or check the list of suppliers in Appendix.

To seal handmade paper in plastic, first separate one sheet of plastic from its backing. Place the plastic sheet on a flat surface with the sticky side up. Be very careful not to touch the adhesive surface as this could leave marks.

Now take a completely dry sheet of handmade paper and hold it so that one edge touches the adhesive of the plastic (see illustration). Hold the other end up high so that there is no danger of accidental sticking. Now carefully press the sheet down on the plastic.

Separate a second sheet of clear, self-sealing plastic from its backing. Set this second sheet on top of the handmade paper in the same manner as was used to lay down the paper. Working from the center out, smooth the laminated sheet so that no air bubbles remain. If necessary, cut away any excess plastic with sharp scissors.

Hold the sheet of handmade paper so that one edge touches the adhesive of the plastic. Carefully lay the paper on top of the plastic sheet.

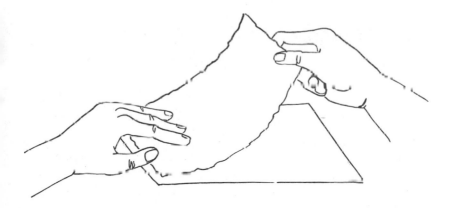

You can make oval placemats by cutting the finished mats to size, using a purchased oval mat as a guide.

Method Two. Many printing shops or office supply stores offer a laminating service. Check the Yellow Pages of your phone book for stores near you. The laminating is usually offered in a variety of sizes and is not expensive.

Method Three. You can also make placemats by inserting your handmade paper into clear plastic placemat sleeves that allow you to change the paper sheets used whenever you wish.

These sleeves are waterproof, foam-backed envelopes 11" x 17" in size. There is an opening in the back. Simply insert a sheet of paper through this opening into the pocket and your placemat is ready to use. See the list of suppliers in Appendix for sources of these sleeves.

Note: You can also use self-adhesive plastic sheets to make other waterproof paper items. For example, handmade paper laminated between sheets of plastic can be cut into coasters to use with glasses. You can even make your own deck of playing cards, painting or printing the faces of the cards and decorating the backs, then sandwiching the paper between two layers of plastic.

PRINTING HANDMADE PAPER

One of the great virtues of handmade paper is that it takes printing inks beautifully. Artists interested in printmaking are drawn to the craft of papermaking to get paper of just the right quality, color and texture for their work. If you are just getting started in printmaking, there are several easy methods you can try that do not require a press.

Use water leaf paper for all your printing projects. You can use either a rough or smooth finish, with different results from each. Remember, the side of the paper formed next to the screen will give a rough texture; the other side will be smoother. Try printing on each side to see which you prefer.

Printing from Objects

One of the simplest ways of making a print is to use an object such as a key, coin or even a leaf and press it into a stamp pad or palette of ink, then press the inked object onto a piece of paper. As you can see, there is very little equipment necessary for this type of printing so it is ideal for children or school groups.

Stamp pads come in several colors other than black: red, green and blue are readily available from stationery and office supply stores. Stamp pads also come in several sizes. Buy a large one if you plan to do much object printing.

You can also use water-soluble printing ink to make object prints. This comes in a wide variety of colors and is sold at art stores. If you use printing ink, you will also need a brayer—a roller with a handle, used to spread the color evenly.

To use printing ink, first spread out a sheet of aluminum foil on a table or counter. Squeeze out a small portion (about a 1" squirt) of printing ink onto the foil. Roll the ink into a smooth, uniformly thick smear using the brayer. The inked area must be slightly larger than the object you plan to print. Take the coin, key or whatever and set it down firmly in the ink. If the ink is not spread too thickly you should be able to pick the item up without getting your fingers inky. Transfer the object to the paper you want to print and press it down firmly. Remove the object and admire your print. You can use multiple prints of the same object or add prints of other items.

With a stamp pad you can also make some attractive nature prints using plant parts. Even if you do not have an herb or flower garden, you can still find many subjects for nature prints: look for shrubs and trees with interesting foliage; check the grocery store for carrot tops and parsley; look closely at the lawn for clover and weeds with nice foliage texture. Use fresh plants if possible since they produce better prints. Pick the plants and use immediately since they may wilt quickly.

Choose a fairly smooth paper surface for making nature prints of plants with fine textures and much detail. This detail would be lost on a rougher surface. Use water leaf handmade paper in any color. You will also need:

> a stamp pad
> refill ink
> tweezers
> plastic wrap
> scrap paper

To make a nature print, first set the leaf or flower that you want to print onto the surface of the ink pad. Place a small piece of plastic wrap on top of the plant part and rub lightly. This will ink the plant while saving your fingers. Lift up one corner of the plastic and remove the plant part with the tweezers.

Carefully set the plant part, inked side down, onto the sheet of paper to be printed. Take a small piece of scrap paper and set that on top of the plant. Rub gently while holding the plant part firmly in place. Do not let the paper slip or your print may smear.

Remove the piece of scrap paper, take off the plant with the tweezers and admire your print.

Single prints can be made with this method but you can also create some beautiful designs using several leaves, flowers and stems. Allow the ink from the first print to dry, then print over it with another plant part. In this way, bouquets can be created on note cards and stationery.

Use different colors of ink, or, for a softer print, apply less pressure. If desired you can repeat a design to make a border.

Stick Printing

This is another simple printing technique that does not require much equipment. You will need:

> small wooden sticks or dowels in different diameters
>   (available at hobby shops and lumber stores)
> a whittling or craft knife

watercolor paints or printing ink
aluminum foil
flat-edged paint brush

To make a stick print, you will first need to cut the ends of the wooden dowels into different geometric shapes. You can cut a different shape on each end of one dowel. Keep one end round for making circles, add a line across another; carve other dowel ends into a square, a triangle, a heart and so on. The shapes can be simple since they will be used in combination to create patterns.

After you have carved the dowels, you can begin printing. Squeeze out a small ribbon of printing ink or mix watercolor paint with a small amount of water to make a thick paste on the aluminum foil palette. Brush the color onto the ends of the dowels. Do not use too much paint or the impression will not be clear. Practice on scrap paper before you use handmade sheets.

To print with the stick, hold the dowel perfectly upright and touch the end of the stick to the paper with even pressure.

Stick printing is good for making borders. Use one or more dowels to create geometric borders on stationery and greeting cards. Build up elaborate designs using several dowels.

You may need to clean the sticks during printing if too much ink or paint builds up on them. Also, clean the sticks before storing.

Potato Printing

This method of making prints is similar to stick printing but the water content of the potato makes the imprint less precise. Materials:

a large potato that is crisp and closely grained
aluminum foil
printing ink
brayer
sharp knife

Cut the potato in half. Using a pencil, draw a design on one of the cut surfaces, then cut away the surrounding potato. To get a clear print, be sure that the design is raised at least ½".

Wipe off the surface of the design with a paper towel to

Handmade paper printed with a potato (left) and a linoleum block (right). Photo by Chris Miller.

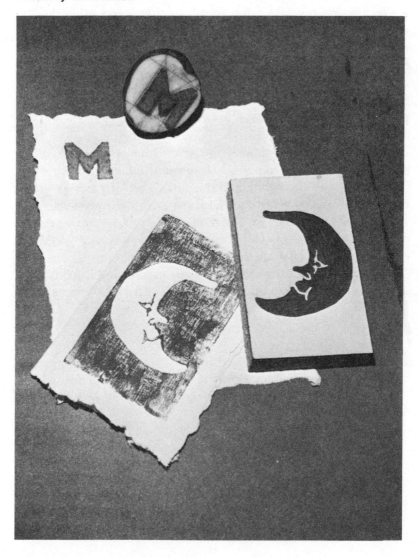

remove some of the moisture. Next spread a small quantity of ink on a sheet of aluminum foil. Roll the ink with the brayer to get a uniformly thick layer.

Press the cut end of the potato firmly into the ink and make a test print by pressing the inked design onto a piece of scrap paper. Check the design and make any necessary changes in the potato.

You can make several prints from one inking, but the color will become progressively lighter. If you want to print with the same potato later, wrap it in a moist cloth and refrigerate. As the potato dries with age, the design will change slightly. So for a uniform batch of prints, it is best to do all the printing using that potato during one session.

Stencil Printing

Here is another easy printing method that requires very little equipment. If you want to stencil initials onto stationery or cards, you can purchase stencils in many different alphabet styles at office supply stores. To make your own stencils, you will need:

> thin cardboard (the kind used to pack shirts)
> paraffin (available in grocery store canning sections)
> sharp knife or razor blade
> watercolor or acrylic paint or printing ink
> brayer or flat-edged brush
> sandpaper
> aluminum foil

The cardboard you use to make a stencil must be thin enough so that the brayer or brush can reach the paper being printed. To waterproof the cardboard, heat the paraffin in a large kettle or double boiler until it is liquefied, then carefully dip the cardboard into the hot paraffin. This waterproofing will keep the cardboard from absorbing ink or paint and it also makes cleanup easier.

To make the stencil, draw a design or initial on the cardboard. Working on top of a stack of newspapers, cut out the design with a sharp knife or razor blade. Smooth the edges of the design with sandpaper if necessary. To make a print, set the stencil on top of a sheet of paper. Spread paint or printing ink on a sheet of aluminum foil and roll the brayer in the color until it is evenly coated or load the brush with paint. Using either brush or brayer, spread the color over the opening of the stencil until you get the desired intensity. Lift up the stencil and check the finished print.

Linoleum Block Printing

This is an easy and inexpensive method of printing. It allows you to get much more detail into your prints than you can with the potato or stick printing methods. You will need the following:

> linoleum blocks in various sizes (available from art and crafts stores)
>
> brayer (choose a soft-rubber brayer about 6″ long and 1″ in diameter)
>
> ink (Either oil-base or water-base printing ink will produce attractive results, but the water-base ink is easier to clean up. Relatively undiluted tempera or watercolors can also be used.)
>
> knives (any sharp craft knife; you will need at least a V-shaped gouge and a straight blade)
>
> metal or wooden spoon

To make a block print, first sketch out a design on a plain piece of paper. When you have a sketch that you like, transfer it to the linoleum side of the block by placing carbon paper between the drawing and the block, then lightly tracing over the design using a dull pencil.

Any part of the design to appear in color must be raised on the finished block; any part not to show up must be cut away.

When cutting out a design, always cut away from your body. Then, even if the knife slips, you will not be hurt. Also, always hold the linoleum block so that your hands are behind the point of the knife blade. Keep the knife sharp since a dull edge is more likely to slip. Cut deep into the linoleum surface. Shallow cuts will fill with ink after just a few printings.

When the block is completely cut, make a test print on a piece of scrap paper to see how the image transfers. Make any necessary changes in the design.

To print from the linoleum block: First cover the counter or table with newspapers. Use a piece of aluminum foil or a shallow aluminum tray as a palette. Squeeze out a small quantity of ink (about a 1″ squirt) and roll the brayer through the ink several times to get an even coating. Then lightly roll the brayer over the surface of the block. Set the brayer back on the palette and carefully turn the block over, inked side down, onto the sheet of paper to be printed. Press the linoleum block against the paper, being careful not to let it slip or the print will be smeared. Then carefully pull the block away from the paper and check the printed image.

Before making another print, ink the block again.

Wood Block Printing

The method used for creating a design on a block of wood is the same as that used for the linoleum block except that the wood is usually clamped to a workbench during cutting. Always cut away from your body, turning the wood in the clamp before making cuts in a new direction.

Choose a soft wood such as pine for the block and use a piece about 1″ thick in the desired size—available from your local lumber store.

To print from a wood block: First set the block, design side up, on a newspaper-covered surface. Ink the brayer and roll it over the surface of the design. Roll the ink on in several different directions. Wood may absorb some color, so additional inking may be necessary. Coat the design with ink until the raised areas appear shiny.

Next, carefully set a piece of paper on top of the wood block and smooth it out from the center. With a metal or wooden spoon, begin rubbing the paper in the center, using a circular motion. Work out toward the edges. Check the print by pulling back just one corner of the paper. If the image is not as dark as you would like, carefully replace the paper and continue rubbing with the spoon. When the printed image is the way you want it, carefully remove the paper by pulling back one corner and peeling the sheet off the wood block.

Wood block prints look different from those made from linoleum blocks. One reason is that the wood grain becomes an integral part of the design. Linoleum blocks are often easier to work in great detail while wood blocks lend themselves to dramatic designs.

Handmade paper is, of course, also well suited to other types of printing: silk screen, lithograph and intaglio. All these methods require special equipment, though, and will not be described here. The Appendix lists books on these subjects if you want to learn more about the possibilities of fine print-making.

10

The Home Paper Shop

If you are interested in making larger sheets of paper than your kitchen sink allows or if you want to make rag paper or experiment with other fiber sources, you will need more specialized, heavier-duty equipment than just a blender, iron and oven.

The main limiting piece of equipment for the home paper-maker is the beater. Beaters are expensive and many craftspeople choose to either buy a used beater or to convert a washing machine to this purpose. Another alternative is to form a co-op with several other papermakers and invest in a beater together. You can also check with local schools and colleges to see what equipment is available in their art departments or contact art associations to locate other craftspeople in your area who may have a beater to rent.

Once you have located a beater, you will also need a vat. This could be half a barrel (check with building supply stores) or a large utility tub. Set the vat at working height to avoid constant stooping.

A press is another necessary piece of equipment if you plan to make paper on a large scale. Printer's presses, art presses and bookbinder's presses can all be used. See the list of suppliers in Appendix for sources.

You will need large felts (see Appendix) and a mold and deckle. Make your own or purchase these from a paper mill or supplier.

Papermakers get their pulp from various sources. You can buy cotton linters or half-stuff. Cotton linters are made of the seed hair of the cotton plant. Half-stuff is new 100 percent cotton fabric which has been cleaned and torn nearly to threads. It is partially beaten and sold by the pound, either damp or dried. Both linters and half-stuff require further beating.

To make rag pulp, clean cotton rags are cut into small pieces (about 1″ x 2″) and beaten together with pH neutral water. The length of beating time affects the finished paper and is an important aspect of the craft. Rag pulp is ready to use when no lumps or threads are visible.

Beating shortens the fibers of rags, fraying them to increase their surface area and swelling action. All this results in a softened, more elastic fiber. Closer contact between fibers is then possible and a better sheet of paper can be formed.

Couching. The couching described in Chapter 5 is a simplified version of the traditional method. If you are making large sheets or a large quantity of paper, use the traditional couching techniques.

Form a sheet of paper, remove the deckle and let mold drain. Dampen a sheet of felt and spread it out on a flat surface. Set one edge of the mold against the felt, then lay the mold on top of the felt and pick up the mold from the opposite edge. The paper sheet should adhere to the damp felt. Place another felt on top and continue your papermaking.

If the paper does not adhere to the felt or if it tears during couching, then try gently pressing on the back of the screen when the mold is turned over on top of the felt. Often this pressure is enough to get the sheet to drop to the felt without ripping. Successful couching requires a good bit of practice.

Pressing. Use a printer's, artist's or bookbinder's press. The

THE COMPLETE BOOK OF HANDCRAFTED PAPER

first pressing given to the stack of wet paper sheets and felts serves to drain out most of the water and it also unites the fibers. After this pressing, the paper sheets are removed from the wet felts and stacked between dry ones. The stack is then pressed again. Finish drying the sheets of paper in the oven or by air drying.

Using these advanced methods with the techniques described in earlier chapters, you will be able to increase your rate of production and the quality of the paper you make.

THE PAPERMAKER'S RECORD BOOK

If you plan to do much papermaking, it might be helpful to take a tip from hobbyist dyers who set up a record book of their recipes with samples of the dyed material.

To make such a record book of your papermaking projects, choose a loose-leaf folder. You can use lined paper and make entries by hand or use bond paper and type the information, adding holes later with a punch.

For each batch of pulp you make, you will want to record:

1. The date.
2. Type of pulp. If recycled, what kind of paper was used? (Noting the colors, amounts of printing and paper finishes might be helpful.) You should also record if rag paper was included and in what amount. Some papermakers may like to measure or weigh the recycled material.

If you are processing your own rags, linters or half-stuff, list the amounts, pH and source.

3. Beating. Record the beating time for rags, half-stuff or linters.

4. Boiling. For recycled paper, note the length of boiling time.

5. Sizing. Any internal sizing added to the pulp should be listed with the amounts used. In this way, you can duplicate a finish later.

6. Additions. Make a note of any extras you add to the pulp:

169

talcum powder, whiting, vegetable fiber, thread and so on. This takes the guesswork out of trying to reproduce a sheet of paper later.

7. Dyes. The amounts of dye used should be written down together with the color and brand names.

8. Drying. Note how the paper was dried: air, oven or iron drying. Record the drying time.

9. External Sizing. If the sheets of paper were sized after being dried, mark this in the record book. Make a note of the type of sizing and whether the paper was C1S or C2S.

10. Dimensions. The size of the finished sheet.

11. Batch quantity. The number of sheets made at this time.

12. Additional notes. You may want to record with what media the paper was used: watercolors, acrylics, etc., and how it behaved. If you sell your paper, you could also make a notation of the buyer. This information could be useful later for repeat customers who cannot remember what paper they bought. Any special problems encountered during papermaking should also be noted here.

After this information is recorded in the notebook, make holes in a sample of the paper with a punch and add it to the notebook. A paper sample glued to a page may discolor or warp because of the gluing, with the result that it is difficult to tell the original characteristics of the sheet.

The record book serves many purposes. First, it enables the busy papermaker to replicate any sheet of paper at a later date. There is perhaps nothing more frustrating than remembering a certain paper but being unable to duplicate it in subsequent paper batches. Referring to the record book will eliminate this problem as much as is possible.

Second, the record book can be a handy selling tool. It can be used as a sample book to show prospective buyers what type of work you have done. Seeing a variety of papers will help customers either to pick out one that suits them or to describe something different.

A record book is also fun for the craftsperson. It will remind you of old recipes you meant to try again. It is also a pleasure to look through the book as a record of where you have been with the craft and how far you have come. You will be able to see your progress with papermaking as your skill has improved.

SELLING YOUR PAPER

There is a market for handmade paper whether you sell to local artists or to craft stores, whether you sell sheets of paper for artwork or make paper into stationery, cards and calendars. In fact, handmade paper is so unique that it easily finds a market and once people learn how much fun it is to work with handmade paper, they will surely come back for more of your product.

Where to Sell. To begin selling your paper, you will have to let people know about you and your work. Offer to speak before various clubs and groups, demonstrating the craft. Often your local newspaper will be interested in your unusual avocation. Call the editor of the section that runs stories about local people and their hobbies. Let the editor know about your papermaking. Be sure any newspaper article states that you sell your work, do custom orders and that you are available for lectures If you teach classes, this should appear in the story too.

You can sell your paper through local stores. Here a business card is effective. Have cards printed on your own handmade paper if possible, listing your name, address and phone number, then possibly "Hand Papermaker," "Custom Work," "Classes"—whatever describes the products and services you offer.

Contact art stores and crafts shops in your area. When you stop by a store, be sure to leave your card. Let them know you are willing to do custom work. Show the store manager samples of your paper and offer to leave your work on consignment if the manager is not willing to buy it outright.

If you do leave your paper products on consignment, let the store manager know that it is your policy to check back in 4 to 6 weeks to replace any items that are not selling well with something different. You can also restock those papers that have been sold during this time.

Should you plan to sell most of your paper to artists, get to know the art community in your city. Contact local art associations, crafts groups, art schools and the art departments of high schools and colleges. Let them know that you make handcrafted paper, show samples of your work and leave a business card. Find out about the activities of art associations and become a member if possible.

If you are making paper products such as stationery, greeting cards, placemats and calendars, then check with local gift shops, crafts shops and department stores to see if they are interested in carrying your work on consignment. Do not overlook the gift shops that are often included in hotel and motel lobbies or at exhibits, museums and historical sites.

Art exhibits and crafts festivals also offer good possibilities for getting to know other craftspeople and artists in your area and for selling your stationery and paper. Be sure to take a supply of business cards with you when attending such an exhibit. Pass these out to artists who may be interested in your paper or put out a supply of business cards at your booth if you are an exhibitor.

If you do enter a show with your own work, consider demonstrating the craft of papermaking at your booth. Sales usually increase when there is a demonstration. Blend pulp ahead of time and take it to the show in large containers. Use a washtub as a vat and air dry the finished sheets of paper.

Mail-order selling is another possible outlet for your work. It is easy to place an advertisement even in large, national magazines. Decide first which market is best. If you are selling paper for artwork, then check artists' magazines and journals for those that carry classified advertising. There is usually an address listed in the classified section where you can write for rates and copy deadlines. If you are selling handmade stationery, check women's general magazines and home-oriented publications.

172

Before you write your ad, you must decide how you are going to send the orders out to customers. Make up a sample envelope or carton containing your work. Weigh the package and determine the cost of postage, checking with the post office for various rates. You will have to either add postage into the total cost of your product or list it separately in the ad (for example, "plus $1.50 postage and handling"). Remember to include the cost of your shipping materials too: the envelope, package, tape and so on.

Writing a classified ad is not difficult. You need to include a description of your product, its price, shipping cost and your address. Many mail-order businesses feel it is best to advertise just one item in an ad, rather than listing several things. Then, when you receive an order or inquiry, you can send out a price list of your complete line of products.

Make a record of all the inquiries and orders you receive. These names and addresses form the start of your mailing list. When you add a new product, send out price lists to these people.

Pricing. Pricing is often difficult but if you check catalogs and price lists from art supply stores and other hand paper-makers to see what they charge for similar items, you will have a general guideline. If you are selling stationery products, browse through local gift shops and card stores to see how these items are priced.

Write down the prices you decide on and make up a price list. It is also a good idea to have prices in mind for custom work, so that when you are asked about special paper projects, you can supply figures immediately.

Custom Work. As you meet artists and other craftspeople, you will probably receive requests for custom-made paper. This is one of the advantages of handmade paper—the paper can be adjusted to the individual needs of the artist.

You will enjoy filling your custom orders if you follow some simple suggestions. First, make sure you really communicate with your customer about his or her paper needs. Try to determine exactly what the customer wants: sheet size, surface, color, texture, additions and numbers of sheets in the order. Decide if you feel you are capable of producing what the customer wants. If you do not believe you can meet the requirements, it is better to turn down the job or to offer alternate ideas.

Second, be specific about price. To avoid any misunderstanding, write out a quotation for the custom job, describing the type of paper you plan to make and giving the number of sheets and the quoted price. Keep a copy of the quotation and give one to your customer.

Then make up a small test batch of the paper, being careful to record the dye, pulp type, beating time—everything you will need later to duplicate the paper. The record book is helpful for this. Take the sample batch to the customer to see if the paper is satisfactory. If it is, you can make up the entire order.

Keep track of your custom orders, preferably in your record book. Write down the date ordered, date delivered and the customer's name and address as well as complete directions for duplicating the paper. Make a note of the intended use of the paper, whether for watercolors, acrylics, pastels or other media. This information could help you later if someone else comes to you looking for a similar paper and will also enable you to easily fill any repeat orders from the original customer.

TEACHING

Another way to earn money with your papermaking is to teach the craft to others. Even though you may still consider yourself a beginner, if you are making paper you can show someone else how to do the same.

Your business card should mention the fact that you teach classes. You can also contact high school and college art departments to let them know that you give special instruction in the craft of papermaking. Art associations and crafts groups are other possibilities. You may even want to contact civic organizations that give classes to see if they would be interested in hiring you as a craft instructor.

Conducting a successful class will require planning and thought on your part. Decide in advance what size class you can comfortably handle. This could be determined by the size of your workroom or the number of molds you have on hand. Remember that a smaller class usually produces better results, with each student learning more when the teacher has more time to spend with him or her individually.

After you have decided on the class size, think about the number of lessons and their content. Write out an outline for the class. You will undoubtedly make changes in the outline as you actually teach the lessons, so be sure to note any alterations as you go along.

Teaching is a good way to learn and it is always rewarding to show others a new skill.

11

Advanced Techniques

Paper itself has become an exciting medium in the creations of many young artists. One such artist is Helene Trosky of Purchase, New York.

Helene's cast-paper works have been exhibited at the Silvermine Guild in Connecticut; the Wichita, Kansas, Museum of Art; Long Island University, and Manhattanville College. Her work is also part of collections at the Hudson River Museum; Staten Island Institute for Arts and Sciences; Arkansas Museum of Art; Wichita, Kansas, Museum of Art and other public and private collections in the United States and abroad.

Her work combines beautifully textured paper with cast pulp and color to create very expressive works of art.

Helene received a B.A. in fine arts from Manhattanville College. She studied painting with Yasuo Kuniyoshi and Camilo Egas, sculpture with Clara Fasano and printmaking with John Ross and Al Blaustein. She also studied the craft of papermaking with Garner Tullis at the Experimental Print Workshop in California.

Helene came to papermaking and her cast-pulp works through printmaking.

"Chairman of the Board," a cast-paper work done by Helene Trosky in 1978. This work measures 26½" x 24" and was made by casting paper pulp over old wooden slats. Photo courtesy of Helene Trosky.

"I was a printmaker," she says, "doing mostly collagraphs (printed collages) involving a lot of texture, and handmade paper was just a natural evolution from textured prints. Casting the pulp was a further advance."

Helene's works have won awards including the National Association of Women Artists award and, in 1968, the Northern Westchester Juried Award. Her work has been favorably reviewed in articles in *Arts, Art News, France-Amerique* and other publications.

Today, Helene has her own studio in Purchase. How does she create her cast-pulp works? To make pulp for her art, Helene uses a variety of materials. She purchases cotton linters from paper companies and recycles cotton fabric and fibers from the clothes dryer. In addition, Helene looks for any materials which might create an interesting texture in her paper: she uses vegetable fibers, woody fibers and even coffee grounds, fur and sand.

Helene does not use any special sizings or other additives when making pulp, except for the textural additions just mentioned. She has experimented with a wide variety of coloring agents: tea, coffee, commercial dyes, acrylics, colored inks, even oils.

To prepare pulp, Helene uses a macerator that has been improvised from an old washing machine. The machine cycle was changed and knives were added to the rotating blades of the washer. This macerator is capable of turning cotton linters into pulp.

For her cast-paper artworks, Helene uses a wide range of mold materials. She has worked with plaster, rubber, wood and sand molds, as well as molds made of synthetic materials. To create unique pieces, Helene makes hand-molded shapes from the pulp or she casts the pulp over found objects that are placed together but not made into a permanent mold.

How does Helene feel about the future of cast paper as an art form?

"I think it has become a permanent medium for artists, but don't think it will ever become a primary art form. I think, like

"Picture at an Exhibition" is a 28½" x 24½" cast-paper work done by Helene Trosky in 1976. The central design was formed by casting the pulp over a hand and the border was formed by casting over an old picture frame. Photo courtesy of Helene Trosky.

all other new developments and techniques, it will become another avenue available to artists for expression and it will be either used alone or combined with other media for maximum effect.

"I think handmade paper is an exciting medium and I still have much exploring to do in and with it. It has so many possibilities, limited only by one's imagination. Right now I am fascinated with the tactile and visual appeal of the texture and color. Naturally this appeal alone has its limitations, but for the foreseeable future I'm hooked!"

MAKING CAST PAPER

Forming paper into decorative shapes is not difficult and you may not even need to make or purchase a mold. Often you can use items found around the house to serve as temporary molds for your paper casting.

Found-Object Molds. Look for bold shapes that will come through to the paper well: seashells, bowls, wooden boxes and so on. If the object you choose is heat proof, then the cast paper can be oven dried.

Purchased Molds. Hobby shops carry a wide variety of molds for candles and all these are suitable for making cast paper. However, if the mold is plastic, do not oven dry the paper or the mold may melt.

Formed Molds. You can create your own molds using plaster of Paris. Add water to the plaster and mix until the material is thick enough to hold a shape as you work with it. On a sheet of aluminum foil, set out the plaster and mold it into the desired shape. Let dry on the foil for a few hours. Plaster of Paris molds can be oven dried with the cast paper.

Note: Some brands of plaster of Paris set up faster than others. Look for a product with maximum drying time.

Casting the Paper. First form a sheet of paper. Be sure to make the sheet large enough to completely cover the mold you want to use for the casting.

Set the mold on a piece of aluminum foil on a baking sheet. Couch the formed paper, then holding the couching cloth by the top two corners, transfer the damp sheet of paper to the mold. Set couching cloth on top of the mold, paper side down, centering if required. Gently pat the couching cloth around the mold. This is especially important if you want too pick up small details. Patting the paper onto the mold will bring out even ridge lines on shells on the finished cast piece. Remove the couching cloth.

Set the cast paper and mold aside to air dry or if the mold is heat proof, place in a 250° F. oven for about 30 minutes. Check after 15 minutes to see if the paper is dry yet. When dry, the cast paper will pull off the mold easily.

Cast paper can be incorporated into artwork such as collages or it can be combined with color and texture.

You can also form pulp over molds by hand. Blend the pulp and pour it into a colander lined with a piece of fiberglass-mesh screen. Let drain, then pick up the screen and squeeze out the excess water. Remove the pulp from the screen and pat over the mold. Dry in an oven at 250° F. or air dry if the mold used is not heat proof. Areas of pulp that overlap on the mold may not bond together unless you add 2 to 3 drops of white glue to each blender-load of the pulp.

By hand forming the pulp, you can create much thicker cast-paper works. You can even combine several different colored or textured pulps in the same molded artwork.

LAMINATED PAPER

Laminated paper offers another area of experimentation for the artist. The double-couched two-tone paper described in Chapter 7 is achieved by lamination. Furthermore, all sorts of items can be sandwiched between two outer sheets of paper: decorative grasses, weeds, dried flowers, leaves, string, postage stamps or photographs. The entire sandwich is then couched and dried together as a unit. The colored watermarks described in Chapter 8 require three sheets of paper to be laminated together.

To make laminated paper, form a sheet and couch it. Set it aside but do not add a felt. Place the objects to be laminated on top of this first sheet. Choose items that are relatively flat or experiment with bulky ones.

Couch a second sheet of paper of the same or a different color. Holding the couching cloth by the top two corners, carefully line up the second sheet of paper with the one that is already couched. Place the second sheet on top. If you want to reveal some of the sandwiched objects, remove the top couching cloth and carefully tear the top sheet of paper above the laminated items.

Air dry the laminated paper if the additions—photographs, fresh plants, etc.—are not heat proof, or oven dry at 250° F. for about 30 minutes. Check often to see if the paper is dry.

PULP SCULPTURE

Pulp that has been prepared in the blender can be molded into various shapes as separate sculptures or you can add them to sheets of paper.

To make pulp for molding, blend 1 tablespoon of boiled paper in the blender with 2 to 3 cups of water and add several drops of white glue. Blend until the paper is broken down into pulp.

Place in the sink a colander lined with a piece of fiberglass screening. Pour into it the pulp and water to drain. Then pick up the screening and squeeze the pulp in it to remove more water. When you open the screen, the pulp should be slightly damp and easily molded into any shape.

Set the finished pulp sculpture on a sheet of aluminum foil to air dry or oven dry the piece at 250° F. Little shrinkage should occur.

If you need more than 1 blender-load of pulp to complete a sculpture, be sure to add white glue to each batch. Pulp used for molding can be colored before it is sculpted or the piece can be painted later.

PULP PAINTING

You can create paintings by adding pulp of different colors and/or textures to a newly-formed paper sheet. Thus, texture, color and design can all be explored with this method.

Pulp can be added to sheets of damp paper to create texture (back) or molded by itself (front). Photo by Chris Miller.

To make a pulp painting, first fill a sink or vat with pulp of the desired background color. Dip a mold and form a sheet of paper.

Set the mold to one side of the sink to drain. You can add other pulp to the sheet freehand, but for more control of the painting, use aluminum foil to make molds.

Use a 2″-wide piece of aluminum foil, folded several times to form a fairly stiff strip. Bend this into the desired shape and staple the ends together. Set the shaped aluminum foil on top of the formed paper sheet on the paper mold. Hold the mold over the sink and pour different colored or textured pulps into the aluminum foil sections. Pulp can be built up quite thick. Add pulp to all the design areas, then let the paper drain over the sink for a few minutes.

184

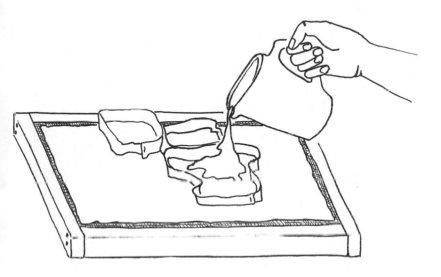

To make a pulp painting, pour colored or textured pulp into the aluminum foil molds.

Carefully remove the aluminum foil molds and check the design. Make any changes by replacing the foil or by adding new sections and then adding more pulp.

Set the entire paper mold on a stack of newspapers and allow the pulp painting to air dry. Change the newspapers in about 15 minutes or so if they become completely soaked. Very thick pulp paintings may take several days to dry. Remove the painting by gently peeling the paper sheet off the mold. When the painting is completely dry, it should come off easily.

Texture can be added to the pulp used for painting: plants, bits of straw, grasses, gold foil, flowers and so on can be incorporated. The possibilities are unlimited and it is interesting to experiment with various additions of texture and color. Finished pulp paintings that are made on small paper molds can be mounted on larger pieces of handmade paper for framing or display.

NONWOODY-PLANT FIBER SOURCES FOR PULP

Any cellulosic material can be made into paper and, since the cell walls of all living plants are composed at least partially of cellulose, that means there is a wide, wide range of potential pulp sources. At one time or another, papermakers have created sheets of paper from almost every plant, from the artichoke to the thistle.

Cellulose makes up about one-third the structure of annual plants and about one-half that of perennials. Cellulose is an organic material made of very strong, durable and flexible fibers—all qualities that make cellulose valuable for papermaking.

The strength of paper is determined by several factors: the strength of the individual fibers in the pulp, the average length of the fibers, the interfiber bonding (which can be increased by beating) and the structure and formation of the sheet.

Cereal straws were used as a pulp source before wood. Straw pulp is still used in some European and Asian countries, particularly those lacking forests.

Nonwoody plants contain less total cellulose, less lignin and more miscellaneous materials. As a consequence, paper made entirely from nonwoody plants tends to be dense, stiff and has a low opacity and a low resistance to tearing. However, if the nonwoody-plant pulp is mixed with other types, such as rag or wood, then these drawbacks can be minimized.

Bagasse is the residue from sugarcane processing and it contains about 65 percent fiber with 25 percent pith cells and 10 percent water-soluble materials. The relatively fine bagasse fibers help make this a satisfactory pulp, especially if much of the pith is removed before processing. Bagasse is used to make paper in countries that have a large sugarcane industry.

Esparto is a grass, native to desert areas of the Mediterranean countries, especially southern Spain and northern Africa. Esparto has a higher cellulose content than is usual for nonwoody plants and therefore makes a good paper. In addition, the fiber size and shape are very uniform.

Esparto was widely used as a pulp source until the mid-1950s when it was replaced by wood pulp. Paper made from esparto forms well and has good opacity and smoothness. In addition, esparto paper is resistant to changes in humidity.

Other nonwoody pulp sources include bamboo, flax, hemp and jute. Some papermakers work with these alternate pulps today, but there are also many plants that are more readily available as pulp sources.

While it is true that few plants will make strong sheets of paper if used alone, it is also true that very interesting results can be obtained by combining pulp from recycled paper (or linters, etc.) with pulp made from garden plants, flowers, shrubs and even vegetables. Use fresh plant material if you want some of the texture of the plants to remain in the paper.

If you want the plants completely pulped, then boil the leaves, stalks and so on for 1 to 2 hours. Use the vent fan if any odor develops. Blend the boiled plants on a high setting of the blender for about 30 seconds. Repeat until no trace of the plants is visible.

Flowers and leaves not only add texture to a sheet of paper, they also add color: pale yellow from marigolds, reds and blues from roses and bright greens from leaves. You can use the dye plants described in Chapter 7 as part of the paper. Simply do not strain out the plant material after the dye bath has been prepared. Instead, blend the plants along with the dyed recycled paper.

Flower Paper. You can add texture as well as color to paper by partially blending flowers and adding them to the pulp.

Try marigolds, daisies, zinnias and chrysanthemums to give a rough, homey texture to paper sheets. Either boil the flowers for 30 minutes to 1 hour or use them freshly picked.

187

Put several complete flowers, including stamens and stalks, into the blender. Add 2 to 3 cups of water and blend for 4 to 5 seconds on a low setting. The flowers should not be completely pulverized, since the charm of paper made from flowers comes from the bits and pieces of petals still remaining.

After the flowers have been blended, stir them into a batch of recycled-paper pulp in the sink or add to the pitcher of pulp to be used with a mold and deckle box in the pouring method. Make sheets of paper in the usual manner.

Petal Paper. To make paper from delicate flowers such as roses, violets, pansies and so on, use just the petals. When you have collected about 1 to 2 cups of loosely packed petals, blend 1 cup of these with 2 to 3 cups of water. Blend on a low setting for 2 to 3 seconds. The petals break down quickly so watch carefully. Add the petals to the recycled-paper pulp when they are in ¼" to ½" pieces.

Vegetable Paper. Herbs and leafy vegetables can also be used to make interesting paper. Try purple basil, mint, sage, parsley and celery.

Cut the vegetables or herbs into small pieces about 1" long and put about 1 cup into the blender. Add 2 to 3 cups of water and blend on a low setting for 2 to 3 seconds.

When the vegetables or herbs are reduced to ¼" to ½" pieces by the blender, they are ready to add to the recycled-paper pulp. Paper formed from the tops of celery has a bumpy texture and may be difficult to couch since the thicker areas tend to tear.

Fragrant herbs also impart their perfume to the finished paper sheets.

Shrub Paper. Use the trimmings from ornamental shrubs, bushes and trees as a pulp source. After branches have been cut, strip off the leaves. These can be added to the blender with 2 to 3 cups of water and blended until reduced to the desired consistency. Or the leaves can be boiled first (recommended for tough, leathery leaves). Most leaves create a bright green pulp, the shade varying with the leaf type and its age.

COLLAGE

Handmade paper is a good choice for collage work and there are two ways it can be used.

Paper can be torn and pasted into a collage in the traditional manner, with other items added or the collage can be created on the mold.

Make several sheets of paper in different colors or with different additions. Couch them and stack between sheets of felt. Make a sheet of paper to serve as the background for the collage and couch it from the mold. Set aside but do not add a felt.

Take out one of the previously made sheets of paper and carefully tear off a piece. Set it on top of the background paper, pressing it in place. Repeat, building up layers of the collage on the damp background sheet. When you are finished with the design, add another couching cloth and iron dry to insure a good bond between all layers of the collage.

COLOR SPRAYING

Delicate, watercolor effects can be created on damp paper sheets using dyes in squirt guns or misters. Use water soluble fabric dye since food coloring fades badly.

Form a sheet of paper by either the vat or pouring method and remove any deckle or deckle box. Set the mold over the sink to drain. Mix one or more strong dye solutions by adding water to the dye powder and fill the misters or squirt guns.

Holding the mister or squirt gun several inches away from the wet sheet of paper, spray on the color. Some misters have adjustable nozzles which allow for a fine or concentrated spray. A fine spray creates a mistlike color on the paper. More concentrated spray makes denser color areas. Be careful not to use a spray that is too strong as it could tear the damp paper. Also, do not use the sprayer or squirt gun too close to the sheet as this can also cause rips.

The color patterns applied with the squirt gun or mister will

Sheets of paper delicately colored with three different methods. Color was sprayed on the left-hand sheet with a plant mister. The center sheet was made by pouring dye from a plastic container with holes punched in it (the Color Clouds method). The paper on the right was sprayed with enamel after it was formed but while still damp. Photo by Chris Miller.

differ from front to back of the paper; choose the design that you prefer.

Watermarks can be highlighted by spraying the dye on the sheet of paper over the watermarked area. Overall patterns can be created or you can add small tint marks or soft borders to paper. Try combining several colors for rainbow paper.

Spray Paint. Interior/exterior spray enamel can also be used to add color patterns to wet sheets of paper. Form a thick sheet of paper. Then shake a can of spray enamel and hold it about 10″ to 12″ from the surface of the paper on the mold. Spray the color onto the paper in one or more spots. Use more than one color or overlap colors. Then couch the sheet in the usual manner and oven dry.

Make different-sized holes in the plastic container about three-quarters of the way from the bottom.

A light concentration of color from the spray can will not disrupt the fibers of the paper. Be careful, though, for holding the spray too close could create weak areas or tears in the paper sheet.

Even though enamel is not water soluble, it still adheres to the damp paper and the color is little affected by oven drying. Light applications of spray enamel do not bleed through to the other side of the paper sheet.

COLOR CLOUDS

Japanese papermakers create beautiful sheets of paper with cloudlike color patterns by pouring dye from a multi-spouted teapot.

191

Since multi-spouted teapots are hard to find, you can make a substitute by using a small, soft-plastic glass, bottle or pitcher. With a nut pick, paring knife, large needle or safety pin, make holes in the container about three-quarters of the way from the bottom. Use several different tools to make a variety of hole sizes. Make all the holes along the same general line; if you make one hole very much lower than the others, dye will flow out of this hole exclusively.

Put about 1 cup of warm water into the container and stir in some dye powder until it is completely dissolved.

Form a sheet of paper in the sink or deckle box and as you raise the paper to the water surface, pour the dye into the water just above the formed sheet. Have another person help if you have trouble keeping the mold level. If the dye is poured onto paper that has been removed from the water, the force of the dye can make holes in the sheet. Even though it is still submerged, the dye can still tear the sheet if you hold the container too close to the paper. Experiment to determine the correct distance for best results.

12

Paper Alternatives

Long before true paper was made, many civilizations were producing paperlike materials from local plant and animal sources.

PAPYRUS

The Egyptians made papyrus from several species of sedge that grew along the Nile River. The most common species was *Cyperus papyrus* which grows 8 feet to 10 feet tall. It has a triangular stalk that is up to 1½ inches wide at the base.

To make papyrus, the Egyptians first cut the stalk into 12" to 18" lengths, then removed the thin rind. Next the pith of the plant was separated into thin strips, the best-quality strips coming from the center of the stalk. Strips made from areas near the rind were used to make a coarse form of papyrus for envelopes or wrapping material. The rind of the plant was discarded.

Working on a board moistened with water, the Egyptian papyrus maker first laid out a vertical layer of pith strips, side by side. When the board was completely filled with this layer or when the desired size had been reached—usually 10" to 15" wide—then a second layer of strips was started at a right angle to the first layer.

Next the papyrus maker either pressed the strips or pounded them with a stone mallet until the layers formed a laminated sheet. No glue was used between the two layers; the sap from the papyrus plant itself served to bind the layers together. In museums, there are papyrus manuscripts dating back to 3000 B.C. and it is interesting to note that although some scrolls have been torn or damaged during the centuries, the two laminated layers of the papyrus have seldom come apart.

After the laminated papyrus mat was formed, it was dried in the sun. Then the surface was smoothed by rubbing with a stone, a piece of shell or ivory. New papyrus was pure white, pliant and very durable.

Each laminated, two-layer sheet of papyrus was joined to other sheets to form a long scroll. The longest scroll known measures 133 feet while the average length is about 50 feet. The type of glue used and the method for joining the sheets to form a scroll is still unknown.

Later the Romans adapted Egyptian papyrus-making techniques, adding a sizing of flour plus a few drops of vinegar to the finished sheets. After the sizing was applied, the surface of the papyrus was beaten smooth.

Papyrus plants have been reintroduced to the Nile area and papyrus is made today at the Papyrus Institute of Cairo, Egypt. The finished product is used in art magazines, for souvenirs and for diplomas at Cairo University. Attractive note cards are made locally by painting the papyrus with tempera.

Papyrus is a unique surface for artwork. The texture and design of the mats are well suited to the use of pen-and-ink as well as tempera.

While papyrus plants are difficult to get in this country in large quantities, there are nurseries that sell some species. (See Appendix.)

J. N. McGovern, emeritus professor at the Department of Forestry, School of Natural Resources, University of Wisconsin, has developed a method for making papyruslike mats from

A greeting card made in Egypt by painting a design on a
mat of papyrus. Collection: John N. McGovern.

A papyruslike mat made from cornstalks by John N.
McGovern.

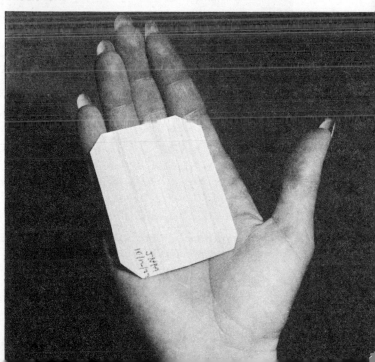

Strips of dried papyrus pith (left) and dried cornstalk pith (right). The cornstalk has the rind still attached. Courtesy of John N. McGovern.

cornstalks, based on the techniques used at the Papyrus Institute of Cairo. Since he first perfected this process in 1972–73, Dr. McGovern has also made mats from the pith of many other plants, including sugarcane, sunflower, hollyhock, canna lily, gladiolus, celery, Chinese cabbage, rhubarb, burdock, mullein, peony, dieffenbachia, dandelion, cattail and thistle.

Depending on the plant material used, the finished sheets of "papyrus" differ in texture, color and flexibility. Sheets made from cornstalks and sugarcane come closest to matching the quality of true papyrus. The main problem with both these plants is that they have nodes on the stem which limit the length of the pith strips.

Using the directions worked out by Dr. McGovern, many pithy-stalked plants can be used to make papyruslike mats.

CORNSTALK PAPYRUS

First cut the cornstalks or other pithy stalks into 4" to 6" lengths with a knife or a band saw. The length should be governed by the pressing facilities available.

Next slice the stalks lengthwise into strips about .04" thick, using the knife or band saw. Separate the pith from the rind with knife or scissors. Soak the pith strips in water for several hours if they are green; soak overnight if the pith is dry. Pith deteriorates if it is stored green, so it must be dried before storing and rewet when you are ready to use it.

Place the pith strips on a wooden cutting board or breadboard and press about ten times with a wooden rolling pin. Rewet the strips and press several more times. The rolling breaks down the pith somewhat and can be compared to the beating step used in papermaking.

Now set the layer of wet pith strips onto a blotter or piece of cloth. Lay these strips in a horizontal row, overlapping slightly. (See illustration.) Press in place with your fingers. Set the second layer of wet pith strips at a right angle to the first layer, forming a cross-laminated mat. Cover the finished mat with another blotter or piece of cloth.

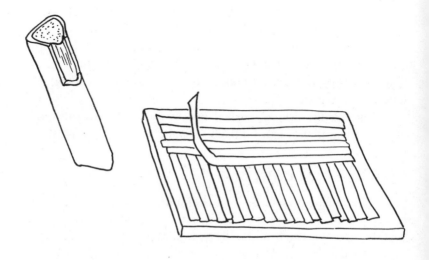

(Right side) Make papyrus mat by setting the second layer of pith strips on top of the first layer. (Left side) Cross section of a stalk of papyrus, showing triangular shape and pith in center.

Place the finished papyrus with the blotters or cloths between two boards. Stand on the boards for several minutes to compress the pith strips and to force out water. Or use a bookbinding or block printing press if one is available. A stack of heavy books can also be used.

Change the blotters or cloths for dry ones and press again. Repeat this exchange until the mat is dry. If you are drying the papyrus mat under books, exchange the blotters or cloths for dry ones and leave the mat overnight or longer to dry.

When the papyrus sheet is dry, a smooth finish can be obtained by running a cold iron across the surface. This step is similar to the burnishing Egyptians did with shells or ivory.

If the papyrus mat curls, it is not completely dry. Place the sheet between blotters again and press overnight.

Papyrus mats made from cornstalks can vary widely in appearance depending on the time the stalks are cut, length of storage and other factors. Some of the mats have a woody appearance while others are fine grained.

PARCHMENT

Parchment is a writing material made from the skins of sheep, goats and pigs. It is believed that a form of parchment was used as early as 1500 B.C., though the credit for its invention is given to Eumenes II, king of Pergamum in the second century B.C. Supposedly, Eumenes developed the process for making parchment because papyrus was so scarce. By the third century A.D., parchment had replaced papyrus as the main writing medium in the Roman Empire.

To make parchment, animal skins were first steeped in pits with lime to remove the hair. Then the skins were stretched onto frames and scraped. Next the skins were sprinkled with chalk and rubbed and polished with stones until they were soft and smooth. The finished skins were then dried until they were about half their original thickness.

Parchment is still available today (see the list of suppliers in Appendix) though it is no longer widely used for diplomas as it once was.

VELLUM

Vellum is considered a very fine grade of parchment. It is made in a similar manner, but the skins of young animals, especially calves, are used.

HUUN AND AMATL PAPER

Around 500 A.D. the Mayans invented a type of writing material made from the bark of the fig tree (*Ficus* sp.). The fig belongs to the same botanical family, Moraceae, as the mulberry tree used by early Chinese papermakers.

Der Permennter.

Ich kauff Schaffell/ Böck/ vñ die Geiß/
Die Fell leg ich denn in die beyß/
Darnach firm ich sie sauber rein/
Spann auff die Ram jeds Fell allein/
Schabs darnach/ mach Permennt darauß/
Mit grosser arbeit in mein Hauß/
Auß ohrn vnd klauwen seud ich Leim/
Das alles verkauff ich daheim.

b   Der

A woodcut from the 1568 edition of *The Book of Trades* showing a parchment maker scraping skins. Courtesy of Dover Publications.

To make their paper, the Mayans first cut branches from the fig tree, some measuring up to 20 feet long. These were then split down the center and the bark peeled off in one sheet. This bark strip was tied into a manageable-sized bundle, then weighted with a rock and soaked for several days in a stream or river. The soaking process was necessary to coagulate the milky sap produced by species of *Ficus* whenever a branch or leaf is cut. After soaking, the coagulated sap could be easily scraped away from the bark.

Mayan men usually handled this first part of the papermaking process, with the women taking over later. It was the women's job to dry the bark in the sun, then soften it with water and beating. The bark strip was laid over a section of log and pounded with a wooden mallet. The pounding separated the fibers of the bark and made them thinner—the end result was a material with a smooth surface that looked as if it had been made in a mold. The pounding process required several hours before the bark was soft, thin and pliable.

The Mayans called their bark paper *huun*. They were the first people to produce "concertinalike" books, made by folding the long strip of bark paper accordion style. These books were then fixed between hard covers.

The Aztecs improved on the Mayan papermaking process, calling their product *amatl*. To create a bark paper with a nonporous surface, the Aztecs used stones called planches which were shaped like flatirons. These were heated, then pressed over the surface of the *amatl* paper to close the pores. This is similar to the glazing process used later by Europeans.

Modern Indians in Mexico make bark paper in a slightly different manner, the resulting sheets closely resembling true paper. After the bark is stripped from a tree, the men use machetes to cut the fiber away. This fiber is then washed and later boiled in a limewater solution. After several hours of boiling, the fiber is removed, washed again and placed in a container.

Indian women fashion paper from the fiber using two tools: a flat board and a beating stone similar to the planche used by the

Aztecs. The papermaker takes pieces of boiled fiber and sets them on the board. Then she pounds these strips with the beating stone until they are felted together to form a sheet of bark paper. The sheet is sun dried on the board. The finished bark paper is smooth on the side that was in contact with the board and rough on the side that was beaten.

Bark paper is used for native paintings and is often sold as souvenirs.

RICE PAPER

This paperlike material is available today as a medium for artwork. It is especially well suited to printmaking.

Rice paper is made in the Orient from the rice paper tree (*Tetrapanax papyriferus*). The stem of this small tree has a large area of white pith, which is cut into thin strings. These are then pressed into sheets of paper.

BIRCH-BARK PAPER

The paper birch (*Betula papyrifera*) grows in Canada, Alaska, and the central and northeast United States. The young trees have a reddish colored trunk, but with age the bark takes on a light color. Trees around ten years old have white or silver outer bark.

Birch bark was used by several North American Indian tribes to make various items: boxes and containers, wigwam covers, canoes, envelopes and scrolls.

Trees were often cut to make stripping off the bark easier, with gathering usually done in the spring. A vertical cut was made down the trunk and the bark peeled away from the tree. Small sheets of bark were stacked on a flat board and pressed with a weight.

No birch tree should be disfigured to obtain bark for your papermaking projects. However, trees that are destined for cutting are good sources of the bark, as is birch firewood.

Stack and weight down the bark sheets. While the bark from the birch tree does not rot easily, mildew is a problem. Check

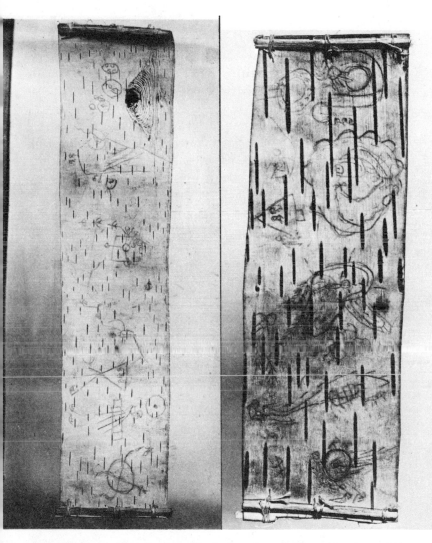

Pictographs on birch bark by North American Indians. Courtesy of the Smithsonian Institution, National Anthropological Archives.

203

the drying bark often for signs of mildew which could create dark blotches. Another problem is that aged bark tends to become brittle.

TAPA

Many cultures around the world, particularly those in Polynesia, have made a writing material known as *tapa,* usually from the pounded bark of the paper mulberry, breadfruit or other trees.

Most often, the tapa was used earlier by these cultures as a clothing material, then later as a drawing and writing material.

To make tapa, islanders beat together layers of the inner bark. Unlike felt, tapa or bark cloth was often given painted or impressed decorations.

Appendix

INSTRUCTION: PAPERMAKING CLASSES
To learn more about the craft of papermaking, contact the following schools which offer classes:

Arizona State University
College of Fine Arts
Tempe, AZ 85281
Phone: (602) 965-6536

Extension Division
California College of Arts and Crafts
Broadway at College
Oakland, CA 94618
Phone: (415) 653-8118

School of Fine Arts
Washington University
St. Louis, MO 63130
(Classes in papermaking and also paper casting)

University of Wisconsin
Art Department
6241 Humanities Building
455 North Park Street
Madison, WI 53706
Phone: (608) 262-1660

In addition to these schools, some of the paper suppliers listed here also offer demonstrations and workshops.

SUPPLIERS
Paper
Aiko's
714 North Wabash Avenue
Chicago, IL 60611
(Japanese handmade paper)

Andrews-Nelson-Whitehead
31-10 48th Avenue
Long Island City, NY 11101
(Handmade and mold-made papers)

Botanica Mills
Box 1253
Fort Worth, TX 76101
(Flower and vegetable papers)

Carriage House
Handmade Paper Works
Brookline, MA 02146
(Papyrus and paper made from plants)

Crestwood Paper Co.
315 Hudson Street
New York, NY 10014
(Handmade and mold-made papers)

Susanne Ferres
3603 South McClellan
Seattle, WA 98144
(Handmade paper, papermaking course)

HMP Papers
Woodstock Valley, CT 06282
(Custom paper)

Imago Paper Mill
1333 Wood Street
Oakland, CA 94607
(Handmade rag paper for art, writing and books)

Paper of All Nations
62 Third Avenue
New York, NY 10003
(Handmade paper, tapa cloth, Mexican bark paper, papyrus)

Special Papers, Inc.
RFD #2
West Redding, CT 06896
(French handmade paper)

Twinrocker
RFD #2
Brookston, IN 47923
(Handmade paper, lectures, workshops)

Upper U.S. Papermill
999 Glenway Road
Oregon, WI 53575
(Handmade custom paper, personalized watermarks, lectures,
    workshops)

207

The William Cowley Parchment Works
Newport Pagnell
Bucks, England
(Fine-quality sheepskin parchment and calfskin vellum)

Yasutomo Co.
24 California Street
San Francisco, CA 94111
(Japanese rice paper, cotton watercolor paper from India)

Pulp
Alpha Cellulose Corp.
1000 East Noir Street
Lumberton, NC 28358
(Linters)

Craftool Co.
2323 Reach Road
Williamsport, PA 17701
(Linters, half-stuff, pulp)

Hercules, Inc.
Wilmington, DE 19899
(Cotton linters)

Imago Paper Mill
1333 Wood Street
Oakland, CA 94607
(Pulp, half-stuff)

Twinrocker, Inc.
RFD #2
Brookston, IN 47923
(Linter fiber, rag half-stuff, pulp, custom sizing, coloring)

Sizing
Imago Paper Mill
1333 Wood Street
Oakland, CA 94607

Papermaker's Mill
P.O. Box 77504
San Francisco, CA 94107

Dyes
City Chemical Co.
132 West 22nd Street
New York, NY 10011
(Mordants, other chemicals)

Colonial Textiles
82 Plants Dam Rd.
East Lyme, CT 06333
(Dyestuff)

Dyers Art/Greenberg
Summer address: Ouaquaga, NY 13826
Winter address:
212 East Broadway
New York, NY 10002
(Dyestuff, chemicals, dye workshops)

Molds and Deckles
E. Aimes & Sons
c/o Green Papers
Hayle Mill
Maidstone, Kent, England
(Custom molds)

Craftool Co.
2323 Reach Road
Williamsport, PA 17701

Papermake
433 Fairlawn East
Covington, VA 24426

Felts
Appleton Mills
P.O. Box 438
Appleton, WI 54911

Craftool Co.
2323 Reach Road
Williamsport, PA 17701

Papermake
433 Fairlawn East
Covington, VA 24426

Beaters
Brill Equipment Co.
35-65 Jabez Street
Newark, NJ 07105
(Used equipment)

Broadhead & Garrett Co.
4560 East 71st Street
Cleveland, OH 44105
(New lab equipment)

Craftool Co.
2323 Reach Road
Williamsport, PA 17701
(New lab beaters)

Presses
Craftool Co.
2323 Reach Road
Williamsport, PA 17701

The Kelsey Company
Meridan, CT 06450

Twinrocker
RFD #2
Brookston, IN 47923
(Heavy-duty presses, also a lightweight aluminum press with a
  maximum sheet size of 22" x 30")

Note: Lockwood's Directory of the Paper & Allied Trades
Vance Publishing Corp.
133 East 58th Street
New York, NY 10022
The Buyer's Guide section lists sources of vats, beaters,
  replacement parts (such as beater bars, bedplates, etc.) and
  also sources of used and rebuilt machinery.

Screening Wire
Estey Wire Works
137 West Central Boulevard
Palisades Park, NJ 07650
(Brass mesh)

Papermake
433 Fairlawn East
Covington, VA 24426
(Polyester screening)

Papermakers Mill
P.O. Box 77504
San Francisco, CA 94107
(Watermark wire)

Sinclair Wire Co.
Springfield, MA 01101
(Laid wire)

C. E. Tyler Industrial Products
Mentor, OH 44060
(Brass mesh)

Plants
Carolina Biological Supply Co.
Burlington, NC 27215
(Preserved plants)

Paul Lowe
3321-C Meridan, South
Palm Beach Gardens, FL 33410
(Papyrus—*Cyperus haspan viviparus*)

Merry Gardens
Camden, ME 04843
(Papyrus—*Cyperus alternifolius*)

Miscellaneous
Basic Crafts Co.
1201 Broadway
New York, NY 10001
(Bookbinding supplies, handmade French marbled endpapers)

The Butterfly Co.
51-17 Rockaway Beach Boulevard
Far Rockaway, NY 11691
(Butterflies)

Complete Scientific
P.O. Box 307
Round Lake, IL 60073
(Butterflies, moths)

Dennison Manufacturing Company
Framingham, MA 01701
(Clear, self-sealing plastic sheets)

Russell's Wallcoverings, Ltd.
1046 Madison Avenue
New York, NY 10021
(Envelope placemats, 11" x 17")

George Bayntun
Manvers Street
Bath BA1 1JW, England
(Bookbinding service)

Phillips & Page
50 Kensington Church Street
London W8, England
(Brass-rubbing supplies, books)

TOURS OF PAPER MILLS

There are some 6,000 paper plants in 49 states and many
offer tours to groups. For information about nearby plants,
check in the Industrial Directories at your local library

ASSOCIATIONS

American Paper Institute
260 Madison Avenue
New York, NY 10016
(Wall chart available showing route of paper from the East)

The Guild of Book Workers
1059 Third Avenue
New York, NY 10021
Phone: (212) 752-0813
(There are 300 members, both amateur and professional,
   involved in hand book crafts: bookbinding, calligraphy,

illuminating and decorative papermaking. The organization also sponsors exhibits, field trips, lectures and workshops.)

National Paper Trade Association
420 Lexington Avenue
New York, NY 10017

Technical Association of the Pulp and Paper Industry
One Dunwoody Park
Atlanta, GA 30341

MUSEUMS

The Dard Hunter Paper Museum
The Institute of Paper Chemistry
1043 East South River Street
P.O. Box 1039
Appleton, WI 54911
Phone: (414) 734-9251

The Dard Hunter Paper Museum is open to the public, free of charge, from 8:30 A.M. to 4:30 P.M., Monday through Friday. Closed on holidays.

This museum is unique. It includes displays of papermaking molds and materials with sample papers from all over the world. There are also exhibits showing how parchment, wallpapers and watermarks are made, and a model of a German hand paper mill.

A book room where interested visitors may read and study the unique collection of books and periodicals on papermaking is part of the museum.

The Colonial Williamsburg Foundation
Post Office Box C
Williamsburg, VA 23185

The Williamsburg Foundation has a papermaking exhibit as

a seasonal summertime demonstration, in operation from the middle of June through the end of August, six days a week from 9 A.M. to 5 P.M. A reproduction vat, couching rack and paper press are all part of this exhibit.

About 5,000 sheets of paper are made each summer in the course of the demonstration. The Foundation uses the paper in projects related to their print shop during the rest of the year. The papermaking exhibit uses cotton linters as a pulp source and on occasion, cotton and linen rags, when available. The Foundation uses a laboratory beater to prepare the pulp. The molds were made by the Foundation and a current project is to produce paper for a book.

For group tours, contact The Group Visits Department at the address given here.

BOOKS

On Paper

Dard Hunter, *My Life with Paper.* New York: Alfred A. Knopf, Inc., 1958.

Dard Hunter, *Papermaking Through Eighteen Centuries.* New York: William Edwin Rudge, 1930.

Jules Heller, *Papermaking.* New York: Watson-Guptill Publications, 1978.

On Dyeing

Dye Plants and Dyeing, #46

Natural Plant Dyeing, #72

Both available from the Brooklyn Botanic Garden, 1000 Washington Avenue, Brooklyn, New York 11225. Order by name and number. Price: $1.75 each.

On Bookbinding

Henry Gross, *Simplified Bookbinding.* New York: Charles Scribner's Sons, 1976.

Aldren A. Watson, *Hand Bookbinding.* New York: Bell Publishing Company, Inc., 1963.

On Printmaking

Grant Arnold, *Creative Lithography and How to Do It*. New York: Dover Publications.

J. I. Biegeleisen, M. A. Cohn, *Silk Screen Techniques*. New York: Dover Publications.

Francis J. Kafka, *Linoleum Block Printing*. New York: Dover Publications.

On Hand Lettering

Helm Wotzkow, *The Art of Hand Lettering*. New York: Dover Publications.

On Making Greeting Cards

John Carlis, *How to Make Your Own Greeting Cards*. New York: Watson-Guptill Publications, 1968.

H. Joseph Chadwick, *The Greeting Card Writers Handbook*. Cincinnati, Ohio: Writers Digest, 1968.

Index

217